HVAC
TECHNICIAN'S GUIDE

HVAC
TECHNICIAN'S GUIDE

ESSENTIAL
ENGINEERING
INSIGHTS

THOMAS ASK

INDUSTRIAL PRESS, INC.

Industrial Press, Inc.

1 Chestnut Street
South Norwalk, Connecticut 06854
Phone: 203-956-5593
Toll-Free in USA: 888-528-7852
Email: info@industrialpress.com

Author: Thomas Ask
Title: HVAC Technician's Guide: Essential Engineering Insights
Library of Congress Control Number: 2026944236

ISBN (print): 978-0-8311-3706-9
ISBN (ePUB): 978-0-8311-9692-9
ISBN (eMOBI): 978-0-8311-9693-6
ISBN (ePDF): 978-0-8311-9691-2

Publisher/Editorial Director: Judy Bass
Compositor: Patricia Wallenburg, TypeWriting
Copyeditor: Judy Duguid
Proofreader: Alison Shurtz
Cover Designer: Jeff Weeks

None of the author's writing was generated or assisted by AI. All photographs are taken by the author except as indicated. Some graphics are produced or modified by AI.

books.industrialpress.com
ebooks.industrialpress.com
1 2 3 4 5 6 7 8 9 10

For Richard Taylor, a colleague who taught much more than HVAC.

CONTENTS

THREE

Making Cold 27

FOUR

Moving Heat and Cold 51

INTRODUCTION

You're good at what you do. You can tell the sound of a good compressor, and you can read a flame. You know the sights, sounds, and smells of an HVAC system.

Let's move this understanding to the next level.

If you strive for excellence and wish to be the best at your profession, it's good to know engineering concepts and the language that lies behind the important work you do. This book doesn't try to present comprehensive design approaches because that's a much longer story. But this writing is rigorous—it hasn't been dumbed down or taken shortcuts. Try to merge these concepts with your day-to-day practice and you will see that there's a strong connection that's useful.

I can't memorize anything. I can only understand stuff. If you are like me, a bunch of names and terminology is not interesting. It is only when you can put the pieces together in a logical way that complex ideas can rest solidly in your mind. The terminology is useful for talking to others, but the terms alone don't help you understand things.

Perhaps the biggest barriers to getting into the engineering side of HVAC systems are math and vocabulary. In this book we're going to look at engineering concepts and the associated terminology. While we won't look at the math, I'm not going to insult your intelligence either. I would like to help drive these concepts home so they dig deeply into your bones. Then I want to give you some of the specialty words that are used. These words are put in *italics*. Words can sometimes be used as a tool to keep people out of the "club"—and I want you to be in the club. I want you to be clear in how you present yourself, wise in your understanding, and fully engaged in the technical marvels of HVAC systems.

The theory studied here relates to the thermal sciences of thermodynamics, heat transfer, and fluid dynamics. I also present underlying concepts related to electronics and building automation. Be patient. Read it like a big story. I even have some corny stories to try to make ideas come alive in a unique way. Section 1 is the most difficult reading because it introduces the most concepts, so just plow in.

I have a Summary of Worries for most chapters. These aren't intended to be comprehensive—you will have a longer list of worries than what I wrote, but they are things to think about every time you deal with the specific parts of a system. I have also included some Safety Summaries and Design Approaches in the chapters. I realize these categories go beyond the boundaries of HVAC theory and into design and practice. They are just a starting point.

As with the Summary of Worries, the Safety Summaries are not intended to be comprehensive approaches to safety but rather illustrative personal anecdotes. Safety is always the most important consideration in your work. The details of safe installation and servicing are beyond the scope of this writing. However, the knowledge gained from this reading will allow you to appreciate safety regulations and have a more intuitive understanding of potential safety problems that elevates your professional HVAC capability. The Design Approaches are provided as a little introduction into common products and designs.

HVAC
TECHNICIAN'S GUIDE

SECTION 1

COMBUSTION AND REFRIGERATION TO THE AIR HANDLING UNIT

ONE

OVERVIEW OF HVAC

HVAC requires knowledge about all kinds of technology—that's what makes it so interesting. HVAC systems use pipes, ducts, fans, pumps, compressors, fuels, refrigerants, water, advanced control systems, chemicals of all sorts, and so much more. These are very complex systems that need to work together. It takes a wide variety of abilities to be a good HVAC technician. Of course, the typical goal of HVAC systems is to provide human comfort and safety, so if you want to start with that presentation (which is logical), you have to jump to Chapter 7, "Conditioning Air."

Overview of HVAC Systems

Let's take a fast walk through a typical commercial heating and cooling system; then we'll come back and look at it in further detail. While there are an increasing variety of ways to heat and cool spaces, we are starting this discussion with a standard boiler and chiller working through air handlers. We won't stop with these conventional systems, but they give us a good way to consider all the concepts that apply to most of the wide-ranging approaches available.

Basically, HVAC systems create heat or cold. They take that heat and coldness and use it to heat or cool water, which is then used to heat or cool air. The air is blown into a room, and everyone is happy. The diagrams below present these common approaches for commercial heating and cooling. We will break these systems into pieces and explore each section.

Heating

For heating applications, spaces are commonly heated with warm air. The warm air can come directly from a furnace or from a fan blowing over a coil with heated water provided by a boiler. Figure 1.1 shows the sequence of creating warm air in commercial systems.

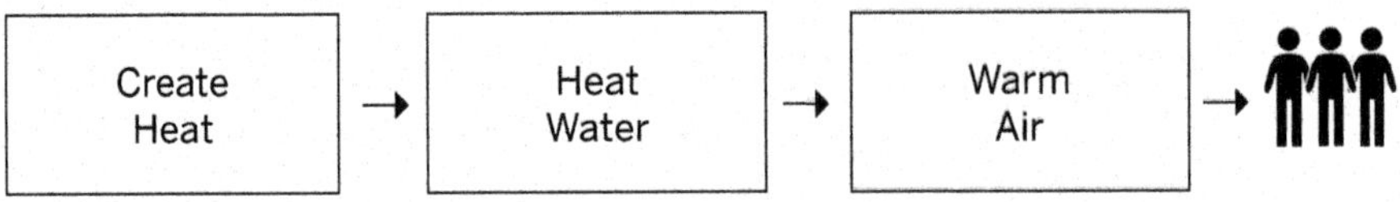

FIGURE 1.1 Stages of heating.

If we transform the block diagram from Figure 1.1 into a simplified schematic, we get the presentation in Figure 1.2. Schematics use standard symbols as shown in the key.

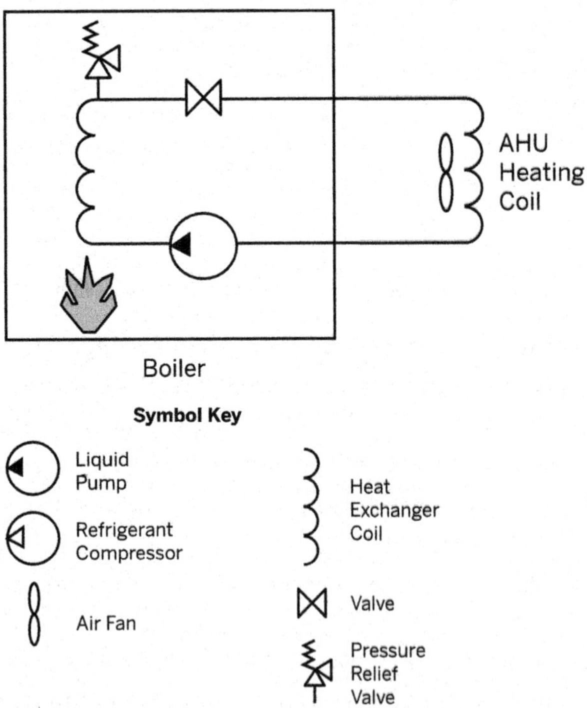

FIGURE 1.2 Simplified hot water heating system with key.

Figure 1.3 labels the parts of the schematic. We will do the same thing for chillers and then break apart the sections as we study them in more detail.

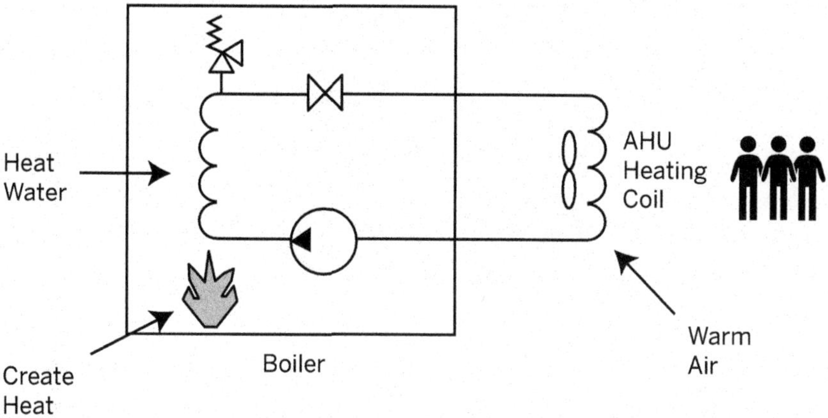

FIGURE 1.3 Stages of heating in a simplified hot water heating system.

Boilers burn some kind of fuel to heat water. The hot exhaust gas is run through a heat exchanger that heats the water that will be distributed in the building. This hot water is pumped from the boiler to air handling units where the water heats air that is sent to rooms and other "conditioned" spaces. The cooled water returns to the boiler. That's basically it. There are a lot of steps involved in getting very hot, dangerous combustion exhaust gases to become pleasantly warm and safe air. A warm air furnace skips the water step and directly heats supply air.

When the hot water gets to the air handler, it runs through another heat exchanger to heat the air. Usually it mixes with some outside air and then gets blown into a room through diffusers or some other terminal device. This supply air does its thing and gets drawn back into the air handler through the return duct. The air can be distributed as a constant air volume (CAV) or a variable air volume (VAV) system. A CAV system pushes a steady supply of air throughout the duct system, whereas the VAV system adjusts the airflow to match the actual load of a zone. These are the basics of a commercial heating system. However, hydronic systems are another way to bypass a heating step. These systems take hot water directly from the boiler to terminal units such as radiators and radiant floor piping. The advantage of this is you have no fan or air movement noise. In addition, if it is

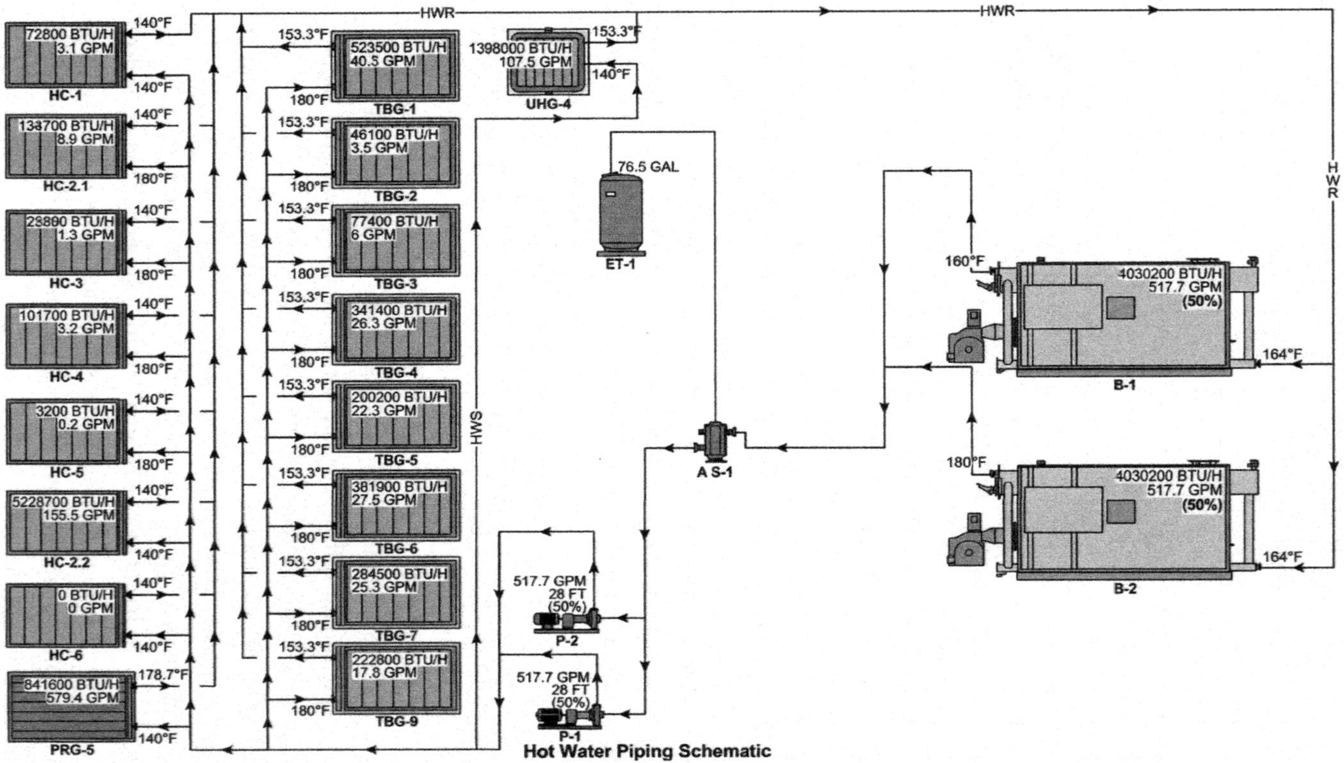

FIGURE 1.4 Schematic of a commercial hydronic system. (*Source:* HVAC Solution)

for heating only, you can avoid ductwork. However, these systems require separate outside air ventilation systems. Hydronic systems can also be used for cooling. In cooling applications the terminal devices are commonly panels or chilled beams. Figure 1.4 shows a schematic of a hydronic system—lots of piping and no ducts.

Cooling

Commercial cooling systems using chillers are more complex in terms of subsystems. Figure 1.5 shows the stages of a chilled water cooling system.

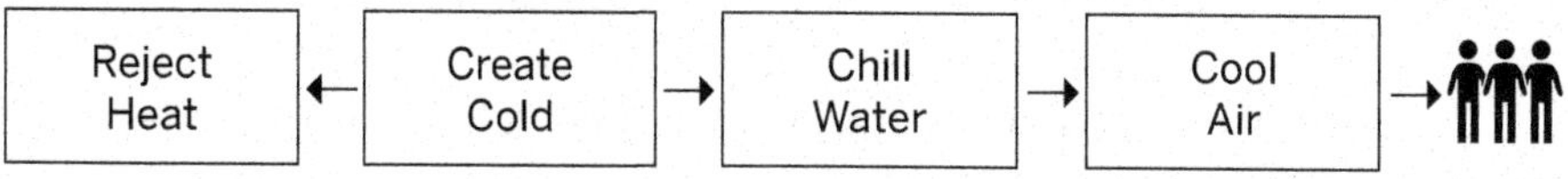

FIGURE 1.5 Stages of cooling.

As we did with the heating system, we can transform the block diagram from Figure 1.5 into a simplified schematic. Figure 1.6 shows the schematic for a water-cooled chiller.

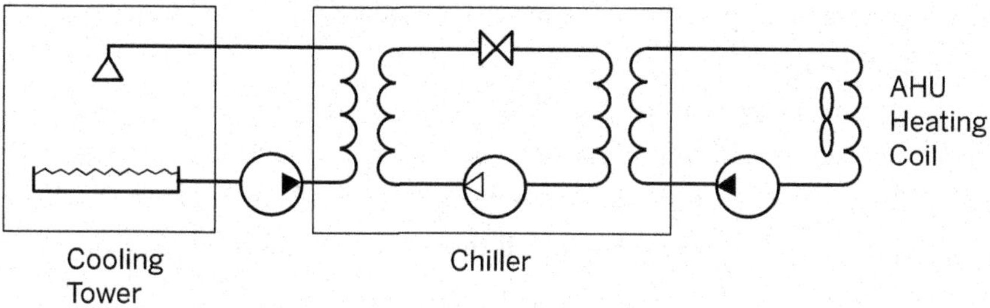

FIGURE 1.6 Simplified chilled water cooling system.

We can label the schematic to understand its function, as shown in Figure 1.7.

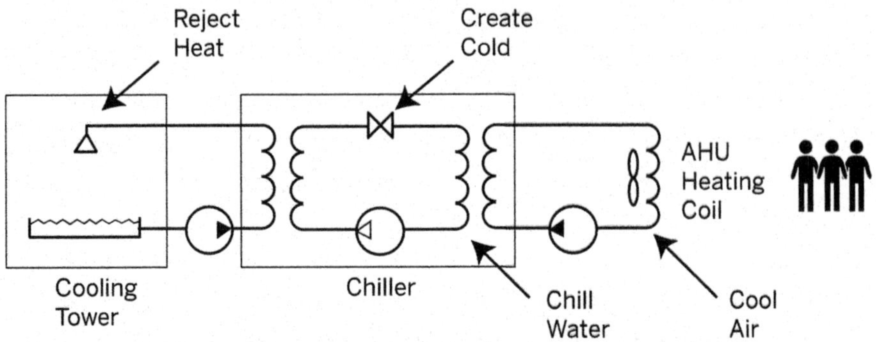

FIGURE 1.7 Stages of cooling in a simplified hot water heating system.

Commercial cooling systems usually use a chiller. Chillers have a vapor compression refrigeration system inside them that converts pressure into temperature changes. These pressure changes to the refrigerant are produced by changes forced upon the refrigerant by a compressor and expansion valve. The cold refrigerant is routed through a heat exchanger (evaporator section) that cools the water that is pumped to air handlers, just like hot water is for heating. The heat produced from cooling in the condenser section is sent through a heat exchanger, and that's usually pumped out to a cooling tower. A typical hot and chilled water system is shown in Figure 1.8.

There are even more steps in cooling than in heating. Here we are converting a dangerous refrigerant into pleasantly cool and safe air. Direct expansion (DX) cooling skips the water step and directly cools air with cold refrigerant. The downside to this cooling system is having all the refrigerant lines running through a building.

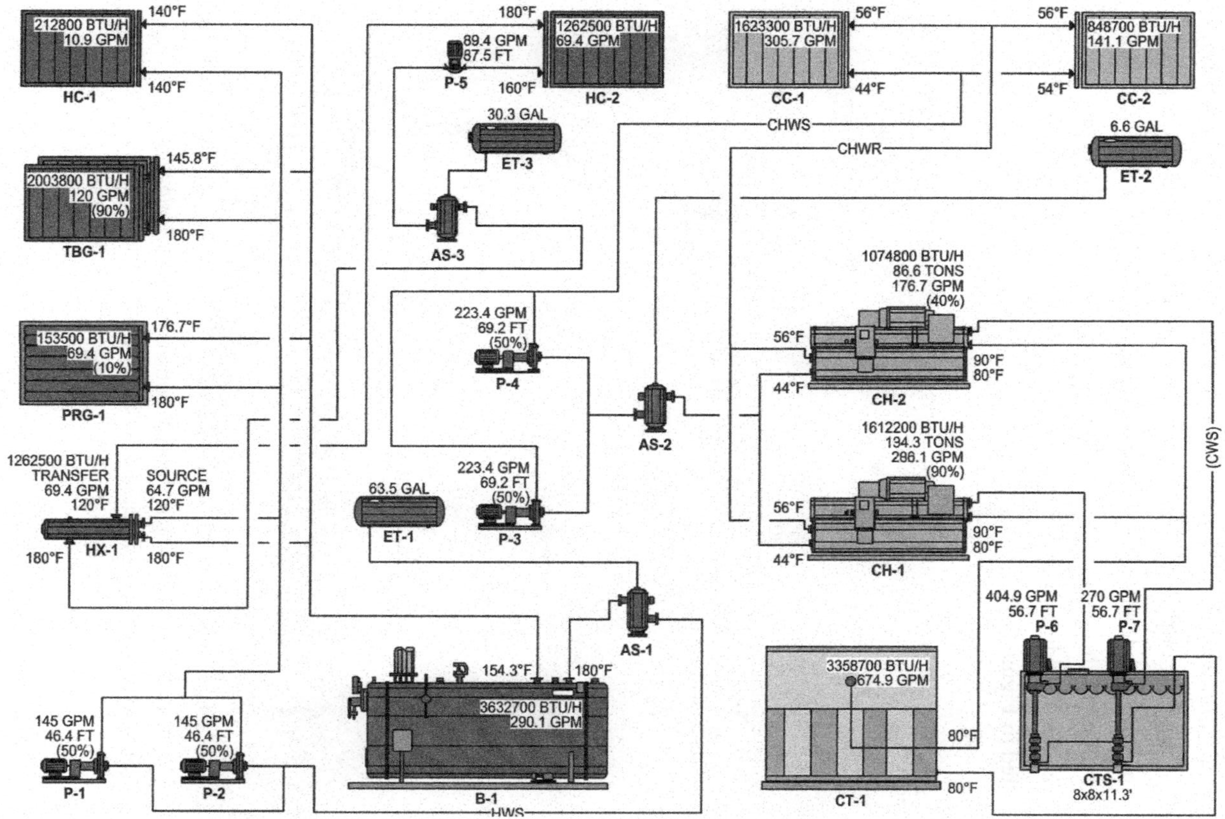

FIGURE 1.8 Schematic of a commercial hot and chilled water system. (*Source:* HVAC Solution)

Makeup Air Systems

Makeup air units are helpful design components because they are an excellent way of preconditioning outside air. These units are dedicated rooftop systems that can heat or precool and dehumidify the incoming outside air. This keeps all that troublesome condensate outside the building. In addition, the units are commonly used to control building pressurization. The crane lift of a rooftop makeup air unit is depicted in Figure 1.9.

FIGURE 1.9 Locating a makeup air unit onto an awaiting curb. (*Source:* Los Angeles Air Force Base Space and Missile System Center, Public domain, via Wikimedia Commons)

Next Step

In the next chapter, we will start digging into the details of heating systems. We will start with the combustion process. Combustion will be followed by fluid flow and finally heat exchange. I'll also mention that there is a discussion of the equations connected with heat transfer and other interesting stuff packed into the Appendix. The Appendix will also include information on thermal load analysis for buildings and homes.

TWO

MAKING HEAT

Heat is usually made with a boiler, furnace, heat pump, or electric resistance. The heat can be distributed by hot water (or steam). Hot water can be routed to air handler coils, radiators, radiant surfaces (such as radiant floors), and other terminal devices. Heat can also be distributed by heating air, as with a warm air furnace.

Boilers

Way back in Chapter 1, I introduced a simple schematic of a boiler. I know you're still thinking it was too simplistic and there is a lot more stuff in a heating system. You're right. But let's pause and look at that simple heating schematic as a starting point for studying the underlying engineering concepts. Figure 2.1 shows the part we will be looking at now. Well, actually we won't be talking about all of this; we will consider heat exchangers and pressure relief valves in a later chapter.

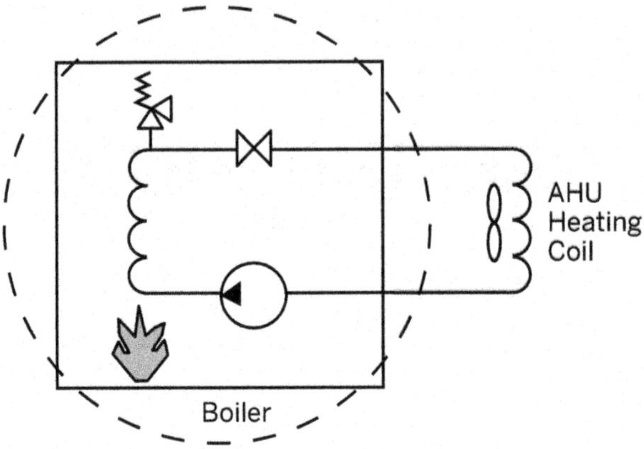

FIGURE 2.1 Boiler section of the heating schematic.

The firetube boilers shown in Figure 2.2 and Figure 2.3 are the workhorses of many commercial heating systems. They burn fuel at one end of a cylinder and run the hot products of combustion through tubes to heat up a water-filled shell.

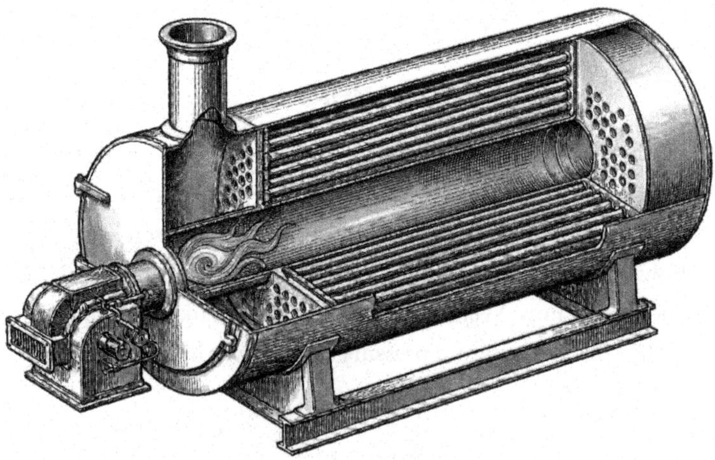

FIGURE 2.2 Firetube boiler.

FIGURE 2.3 Commercial dual-fuel (natural gas and oil) firetube boiler. The large natural gas line dropping in vertically on the left side has many redundant controls as it is piped horizontally to the burner at the right.

Watertube boilers are used for large applications. These boilers have heat exchangers at the discharge of the combustion chamber. Some very large systems will add another heat exchanger to the stack to preheat combustion air. While some boilers create steam, that is not very common anymore. You may think it's funny to still use the word *boiler* when we are making hot water, but it's easier than *high temperature water generator* or all the other terms that rise and fall in usage. There are other heating terms in common usage, but this book is about engineering concepts, not all the HVAC technology and devices out there. We're looking at the theory that underlies them. I'll stick with *boiler*, and I'll also use the word *furnace* for the direct creation of hot supply air. Feel free to replace *furnace* with *warm air furnace* if you are a zealot.

Combustion

Let's look at how heat is produced by fuel. Of course, you recognize that heat can be produced from other sources, most notably electric resistance heaters like in your toaster. This simply requires you to run electrical current through resistant wire. The heat produced is a product of the current and voltage. That's too easy though, so let's look at combustion—it's much more interesting.

Safety and Stack Velocity

The idea of burning huge amounts of natural gas or oil in a metal box should concern you. You are trying to tame a dragon. If it gets out of control, you get a giant fireball or an explosion. You can also create unstable combustion that can produce flashback or blowoff of a flame. On top of the need to tame a fire, combustion creates dangerous gases that can suffocate you and be toxic to the environment. After fuel is burned, the products of the combustion transfer heat to water (or air), and then the products are routed through a stack. You need to have enough pressure and velocity to go from combustion to the stack with enough force to get through any heat exchangers and get launched high enough into the atmosphere to spread out the nasty products of combustion.

Fortunately, there are many features that keep things from getting out of control—for example, a flame sensor, flame rollout switch, limit switches, draft pressure switches. Other sensors are monitoring such things as carbon monoxide, unburned fuel, and condensate. Gas systems have extensive testing and certification requirements. They usually have redundant gas valves that can quickly shut off the gas supply if a malfunction is detected.

Atomizers

Fuel nozzles are designed to fill a combustion chamber while burning enough fuel to get the energy you need. You want to use as much of the combustion chamber as possible without impinging on the refractory. Designing a nozzle is a balance of flame shape and fuel volume. For liquid fuels, droplet size is important, a subject that is expanded upon in the Appendix. The swirl vanes that control the flow movement and fuel mixture are an important part of the burner assembly. Controlling the fuel pressure and nozzle geometry are important for maintaining stable combustion.

For liquid fuels, you develop the fuel angle and hole sizes to push out enough fuel to heat with, while keeping the velocity high enough to atomize well. You shape the hole directions to control the shape of the flame. The hole size dictates the velocity of the fluid and leads to the quality of atomization. The angle and the number of holes give you the flame shape and fuel volume. I designed a nozzle for an experimental fuel for which the "engineering" really became a process of trial and error. I babbled some fancy words and then had a machinist make several different configurations. Then we experimented. Figure 2.4 shows the best design. I guess you could call that research . . .

FIGURE 2.4 Author's nozzle design for experimental fuel.

Fuels

Fuels used in HVAC typically contain hydrogen and carbon, usually natural gas and fuel oil. Combustion occurs when hydrogen is burned with oxygen to produce water vapor or when carbon is burned with oxygen to produce carbon dioxide. The fuel has to be heated to a certain point to start the reaction, just like lighting a candle. This energy is called *activation energy*, and once the reaction starts, the chemically created energy keeps the combustion going, again just like a candle. Specifically, what has happened in these exothermal reactions is that when an activation energy is achieved by an ignition source, such as a pilot light or spark, hydrogen and carbon molecules start reacting with oxygen and produce heat. The term *exothermic* simply means that heat is produced by the reaction and that the final products of the reaction are at a lower energy state. Fuels have other important considerations as well, such as *flash point*, which is the lowest temperature a liquid fuel will produce a vapor. There are other fuel performance specifications, but that's enough for what we are considering here.

Combustion of complex hydrocarbons, like oil, produces short-lived, intermediate products of combustion. That is, the reaction goes through several steps before the final product. This is why reactions require time (as in the ancient understanding of the need for time, temperature, and turbulence for combustion). Carbon monoxide is the most famous, and deadliest, intermediate product of combustion.

Combustion in Air

In practice, most fuels are burned in air, which has lots of nitrogen (there is almost four times as much nitrogen than oxygen in the atmosphere). Nitrogen does nothing but absorb energy, which is why a pure oxygen source (like in an oxyacetylene torch) burns at a much higher temperature than atmospheric air. All that nitrogen in the air steals heat away from the reaction like a sponge because the nitrogen gets heated up to the same temperature as the carbon dioxide and water vapor. If you get rid of the nitrogen, all the exothermic energy goes into fewer molecules. In a similar way, the "stealing" of heat energy is why carbon burns hotter than hydrogen. Hydrogen reactions produce water, which has a very high *specific heat*, meaning it takes a lot of energy to increase its temperature.

Carbon combustion produces only carbon dioxide, which has a much lower specific heat than water. Therefore, for the same amount of heat energy, the carbon dioxide temperature climbs much higher than water vapor. One other problem with combustion in air is that, at very high temperatures, nitrogen will react with oxygen to produce *nitrous oxides* (NOx). *Nitrous oxides* is the group name for nitric oxide (NO) and nitrogen dioxide (NO_2). There are a couple of other small nitrogen-based gases included in this group of reactive gases. Nitrous oxides create nitric acid, which is a contributor to acid rain.

Simple Combustion of Methane

Let's consider the hydrocarbon reaction of methane, which is the primary constituent of natural gas. Methane is the simplest hydrocarbon available; it only has one carbon atom and four hydrogen atoms per methane molecule. To burn methane, you mix it with air and light it up. It produces carbon dioxide, water vapor, and a little bit of nitrous oxide. All the nitrogen in the air just gets heated up; the nitrogen is not part of the reaction. Because inert nitrogen is mixed with the oxygen in atmospheric air, the nitrogen acts as a "sponge" for much of the heat produced by combustion. The nitrogen bounces around in the middle of the combustion process and interacts with the reaction and absorbs heat. The nitrogen can react at high temperatures to produce nitrous oxides. There is a tiny amount of argon and other gases in the atmosphere too, but let's ignore those.

If you write out the reaction of methane in air with all the fancy chemical symbols, it looks like this:

$$CH_4 + (O_2 + N_2)\text{ air} \rightarrow CO_2 + H_2O + N_2 + \text{a little bit of NOx}$$
$$(+\text{ lots of energy})$$

NOx and SOx Pollution

The amount of nitrous oxide produced is related directly to flame temperature. Flame temperature is usually lowered by recirculating exhaust gases back into the flame. Nitrous oxide is also reduced by having it flow through a chemical catalyst that reacts with nitrous oxide. This is what the catalytic converter in your car does.

With more complex fuels, such as coal and wood, other products of combustion arise. Most notable is the combustion of sulfur, which exists in materials like coal and tires. This combustion produces *sulfur oxides* (SO_x). This term describes the collection of these oxides, which consist primarily of sulfur dioxide (SO_2) and sulfur trioxide (SO_3).

Complete Combustion

The reaction of methane shown above is for *complete* combustion. That is, every fuel molecule has just the right number of oxygen molecules to react with, neither too few nor too many. This combustion makes the fuel burn hottest and most efficiently. However, combustion can also be *incomplete*, in which case there is not enough oxygen to match all the fuel molecules. Incomplete combustion will produce either unburned fuel or carbon monoxide. This is why excess air is used in combustion systems. The extra air ensures there are enough molecules of oxygen to burn off all the fuel from its intermediate products, such as carbon monoxide. The disadvantage of extra air is that it reduces efficiency, because all the extra air simply absorbs the heat of reaction and goes up the stack.

Explosions

When the fuel-to-air ratio is perfectly aligned to eliminate either unburned fuel or unused air, the ratio is called a *stoichiometric* ratio. If the fuel is burned without enough oxygen, it's called *sub-stochiometric*. This will prevent all the fuel from being consumed. This has been used advantageously to lower combustion temperature as part of a staged combustion process, where the fuel is burned in a couple of stages. As mentioned previously, the more common fuel and air control introduces excess air to ensure complete combustion of the fuel. Of note, there are certain fuels, hydrogen most notably, that at certain fuel-to-air ratios become explosive. The upper and lower *explosive limits* of these fuels are well known and must be handled especially carefully. Explosions aren't good for the industry (which is why you have ASME pressure relief systems that look like they were designed a hundred years ago).

Table 2.1 shows the lower and upper explosion limits for some gases as concentration percentages in air. Small-diameter solid particles can also be explosive.

TABLE 2.1 Lower and Upper Explosive Limits

Gas	LEL	UEL
Acetone	2.5	13
Butane	1.8	8.4
Methane	5	15
Hydrogen	4	75
Propane	2.1	9.5

Air Pollution

Ash particulates are another important product of combustion. Particulates are usually inorganic materials that can't burn. However, particulates can be made from unburned liquid or solid fuel. Particulates create problems because they can be inhaled deeply into the lung and block the air sacs (alveoli) in the lungs. There is a classification for these tiny particulates, called *PM2.5*, that are *respirable*, meaning they are small enough to be inhaled deeply. Larger particulates are trapped by the mucous membranes in your throat. In addition to the challenges with particulates being inhaled deeply into the lungs, sometimes they carry materials that are toxic, such as arsenic, lead, and mercury. The particulates become a delivery system for these toxic materials, just like cigarette smoke delivers nicotine. There is a special category of air pollution called *air toxics*, which is part of the persistent chemicals category. These are sometimes called *PBT* (persistent, bioaccumulative, and toxic) chemicals but are also described by one of the most extensive collection of acronyms I know, including POP, PFOA, and PFOS. These chemicals are specifically tracked because they are known to cause things such as cancer, birth defects, impaired lung function, and reduced fertility. They can also cause harm to the nervous system, cardiovascular system, and brain.

I'm going to share some corny, fun stories at the end of this section that should help you understand combustion and phase change. I'll do the same with the refrigeration cycle and air handlers.

Types of Fuel

Distillation of crude oil produces distillates with very few sediments. The *distillates* are classified as kerosene—No. 1 and No. 2 fuel oil. The remaining fuels are sometimes called *residual* fuel because they have more thick residues in them. These include No. 4, No. 5, and No. 6 fuel oils. No. 6 (or Bunker C) fuel oil is the cheapest fuel and is used in large commercial boilers. No. 6 fuel oil needs to be preheated to pump. It has more sulfur and other pollutants than No. 2 fuel oil and needs emission controls. There are subcategories of these fuels that relate to their sulfur content. Fuel oil is generally cheaper and easier to transport than gases; it takes about 135 cubic feet of natural gas to equal the energy of 1 gallon of No. 2 fuel oil.

Table 2.2 shows some of the specifications of common fuels; the higher flash points of No. 2 and No. 6 fuel oil make them safer to work with than gasoline with its −45 °F flash point.

TABLE 2.2 Specifications for Common Fuels

	Kerosene	No. 2 Fuel Oil	No. 6 Fuel Oil
Gravity-API	42.8	33.5	8.4
Flash point °F	140°	165°	169°
Viscosity-CS @ 104 °F	1.51	2.85	—
Pour point °F	-65° to -25°	-10° to +5°	+15° to +30°
Sediment & water, vol.	0.001	0.017	0.070
Ash, wt.%	—	0.002	0.053
Sulfur, wt.%	0.05	0.231	2.77
Heat of combustion, BTU/gallon	134,200	138,400	155,500

Table 2.3 provides the energy density of common fuels based upon both volume and weight.

TABLE 2.3 Typical Energy Density for Common Fuels

Fuel Type	Energy Density (BTU/lb)
Hydrogen	61,000
Gasoline	20,000
Kerosene	19,000
No. 2 fuel oil	19,200
No.6 fuel oil	18,200

The energy from burning the fuel comes from breaking the hydrocarbon bonds and reducing the fuel to water and carbon dioxide. The energy produced is equal to the difference between the chemical energy of the fuel and the energy of the resulting water and carbon dioxide. We talked about activation energy before. This is the energy needed to start combustion and is a fancy way of saying you need an igniter. This starts the initial combustion. The energy produced by the combustion is sufficient to continue burning the rest of the fuel. The refractory is designed to keep heat radiating onto the incoming fuel. Some heavy fuels need an ongoing supplementary ignition source to burn well.

The goal of fuel is to deliver energy efficiency. Because HVAC systems are normally stationary, fuel weight is not as big an issue for them as it is for cars and other mobile applications. The energy density of liquid fuels is extremely high compared with many alternatives. Fuel selection varies with time, and the prevailing fuel selection is dependent on many things such as emissions, cost, availability, infrastructure, and, of course, safety. However, some fuels are very volatile, like gasoline, so we deal with it as best we can in our cars.

Stationary heat plants are point source polluters so they can include emission control and can be regulated. Stationary heating systems also have the room and capacity demand to justify preheating, straining, or even polishing fuel as necessary. Polishing fuel is most often used with diesel fuel. It is a recirculating system that runs the fuel continually through filters to remove water and other contaminants.

SAFETY SUMMARY
Potential Explosions

If a boiler loses its flame, you have to go through a purge cycle to make sure any unburned fuel is evacuated from the system. When I've dealt with experimental fuels, this became very important, because some of the fuel could cling to the refractories where it could volatilize and become potentially explosive if the fuel-to-air ratio falls into the explosive range. Always use a long air purge to get rid of fuel vapors. Most of you have probably dealt with an acetylene pop at the least. Imagine that scaled up ...

The term *deflagration* describes the low-speed, subsonic combustion we normally want. *Detonation* describes an explosive combination that travels at high speeds. The speed of the combustion products is so high that they become supersonic and create a wall of dense air that creates havoc. That's not a good thing in HVAC systems. Figure 2.5 illustrates the difference between deflagration and detonation.

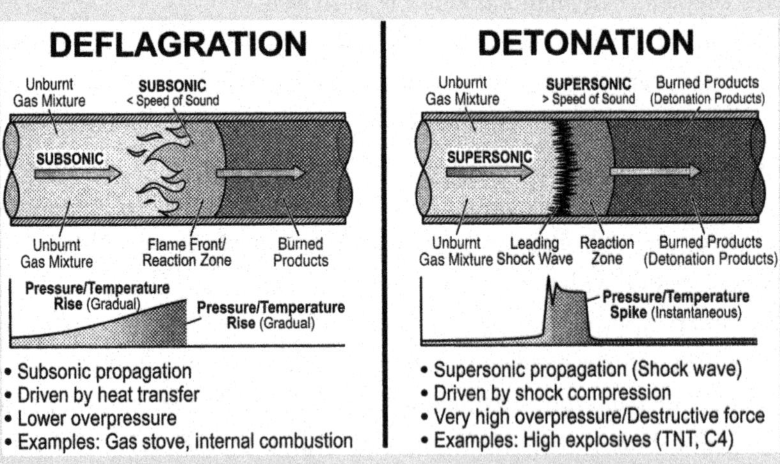

FIGURE 2.5 Deflagration versus detonation.

An inert gas system is used in some systems to prevent explosions. These systems use nitrogen (or flue gases) to eliminate or reduce oxygen concentrations below the explosive limit. Sometimes flue gas is used for this purpose because the product of combustion is largely nitrogen and carbon dioxide and because the flue gas has a very low oxygen concentration.

This is a good time to mention *confined spaces*. These are spaces, like the inside of tanks, that are governed by OSHA because they can be potentially dangerous in insidious ways. These sneaky hazards can include a lurking pool of refrigerant or toxic gases, converging walls, and potential entrapment. Check out OSHA Standard 1910.146 (Permit-required confined spaces) for more information on this. In any event, be careful about anything that looks like a confined space. Yes, that can be part of the job, but you should have a lurking fear.

SUMMARY OF WORRIES

Here are some things you should always be worrying about regarding combustion:

1. Fuels not only burn; they can explode.
2. Carbon monoxide is odorless and colorless.
3. Propane is heavier than air.
4. Fuel can leak. In a confined space it can explode. Under pressure it can atomize.
5. Stack velocity has to be high enough that the products of combustion can escape.

THREE

MAKING COLD

Refrigeration and Chillers

Let's start with the refrigeration section of the cooling system, as shown in Figure 3.1. Large commercial cooling systems use chillers that have a vapor compression refrigeration system buried in them.

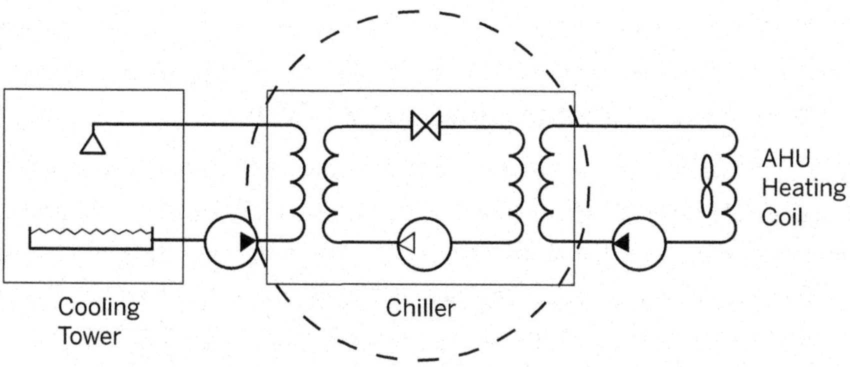

FIGURE 3.1 The chiller section of the cooling schematic.

Unlike DX (direct expansion) systems, chillers send out chilled water instead of refrigerant to all air handlers. Water is safer and cheaper than refrigerants. A chiller operates by using the vapor compression system that you're familiar with. It uses the dance between pressure and temperature to move energy around. A compressor increases pressure, and an expansion valve decreases pressure. This pressure difference equates to a temperature difference that causes heat transfer where you want it—cooling in the conditioned space and heating to the atmosphere. Because you have dealt with this cycle so much, I will try to describe the refrigeration cycle in a unique way with Corny Story 3 in Chapter 6.

Chillers use heat exchangers to send the refrigerant's coldness out from the evaporator to the chilled water that will go throughout a building. Yes, I know about the second law of thermodynamics dictating that energy moves from hot to cold. I just like to make it easy by saying, "sending coldness." And it sounds funny.

Safety Standards

ASHRAE Standard 15 (Safety Standard for Refrigeration Systems) provides information on safe design, installation, and operation of refrigeration systems. ASHRAE Standard 34 (Designation and Safety Classification of Refrigerants) assigns names and safety groups. Regulators oversee much of this industry. This includes the EPA, DOT, OSHA, CPSC (Consumer Product Safety Commission), PHMSA (Pipeline and Hazardous Materials Safety Administration), and local codes.

Chillers Versus Direct Expansion (DX) Systems

While DX systems are a more efficient way of using the coldness of refrigerants, the potential hazard of all that refrigerant running around a building isn't the only problem. Chillers keep the troublesome stuff in the mechanical equipment room, and they can be sized for net demand instead of peak loads. Chilled water can easily be routed to where the peak demand arises. The *diversity factor* is the measure for how much you can reduce the equipment sizing because of the different times that peak demands arise. This is the wise judgment side of HVAC sizing. It's always easy to oversize things—getting it just right is tough.

Most large chilled water systems, as shown in Figure 3.2, use cooling towers to reject the heat. However, there is a shortcut in the form of air-cooled chillers, as shown in Figure 3.3. Air-cooled chillers are a self-contained system that can work in environments that aren't too humid and hot. Packaged units of any type are popular with everyone because they install faster and there are fewer surprises. However, these systems are generally not as efficient as water-cooled chillers. Absorption chillers rely upon heat to create cooling, so these can be attractive if there is a cheap form of heat available. They will be discussed later when we look at thermodynamics.

FIGURE 3.2 Water-cooled chillers are the workhorses of most commercial buildings' cooling systems.

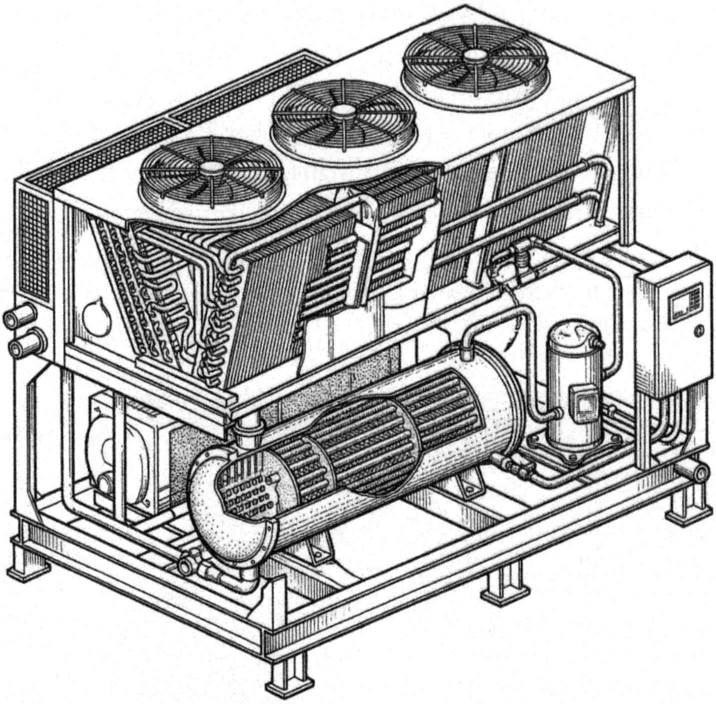

FIGURE 3.3 Air-cooled chiller showing a fan-cooled condenser on top of the assembly.

Refrigeration Cycle

The basic concept of the vapor compression cycle is that forcing a pressure change forces a temperature change. This is why there are compressors and expansion valves in refrigeration systems. Compressor designs will be discussed later; the point here is that compressors raise the pressure of the refrigerant. Expansion valves do the opposite; they lower the refrigerant pressure. The simplest form is a tiny-diameter capillary tube. This design works satisfactorily in small, fixed-capacity systems. The "cap tube" is easy to plug, and we all learned to use a triangular file to cut the tubes. More adaptable designs are the thermostatic expansion valve (TXV) and the electronic expansion valve (EEV). TXVs are not electronic; they use a sensing bulb at the evaporator outlet to send a pressure signal to the TXV diaphragm that controls the position of the valve (Figure 3.4). EEVs use electronic sensors to measure refrigerant temperature and pressure at the evaporator outlet. This signal is sent to an electronic controller that drives a motor in the EEV to position the flow control valve.

Commercial rack systems, as shown in Figure 3.5, use multiple compressors that can be staged as needed. These are commonly used in supermarkets and cold storage facilities.

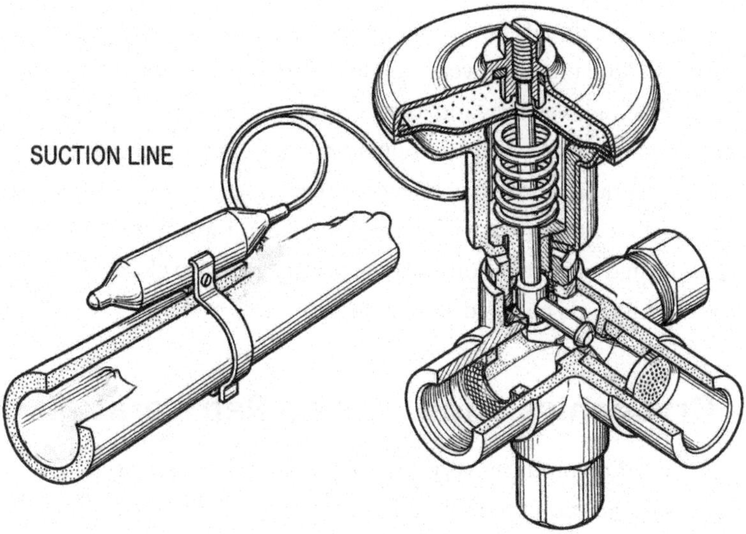

FIGURE 3.4 TXV showing the bulb attachment to the suction line that feeds back to the diaphragm.

FIGURE 3.5 Commercial rack/distributed refrigeration system using DX cooling.

The refrigerant's evaporator is coupled to the chilled water system through a heat exchanger. We discussed how heat exchangers work before. Liquid-to-liquid heat exchangers have the same design considerations, namely surface area, temperature difference, and flow rate. These are the keys to controlling heat transfer. Heat transfer is also affected by something we will talk about later called the *boundary layer*.

While all these details become very important in making heat exchangers work effectively, fundamentally you want heat exchangers to have a high temperature difference, a high velocity, and a lot of contact area. You want materials with high thermal conductivity, such as copper and aluminum. While the evaporator side of the refrigeration system connects to the chilled water, the condenser side of the refrigeration system is run through a refrigerant to a water heat exchanger. The water is heated up by the hot side of the refrigeration cycle and, in the case of water-cooled chillers, is pumped to a cooling tower.

Cryogenics and Cascade Refrigeration

Cascade refrigeration produces very low temperatures. This clever design connects two or more refrigeration cycles together as shown in Figure 3.6 . The low-temperature circuit absorbs heat from the refrigerated space and sends that heat to the high-temperature circuit. Instead of rejecting heat directly to the air, the low-temperature circuit transfers it to an already cold evaporator in the high-temperature circuit. The high-temperature circuit, in turn, transfers all the heat to the atmosphere. This approach lets systems operate well below –140 °F.

Liquid nitrogen is commonly used for extremely low temperature, cryogenic cooling used for such things as medical procedures and environmental chambers. These systems use extremely cold liquid nitrogen (–321 °F) to conductively transfer heat from a location as the nitrogen changes phase from liquid to gaseous states.

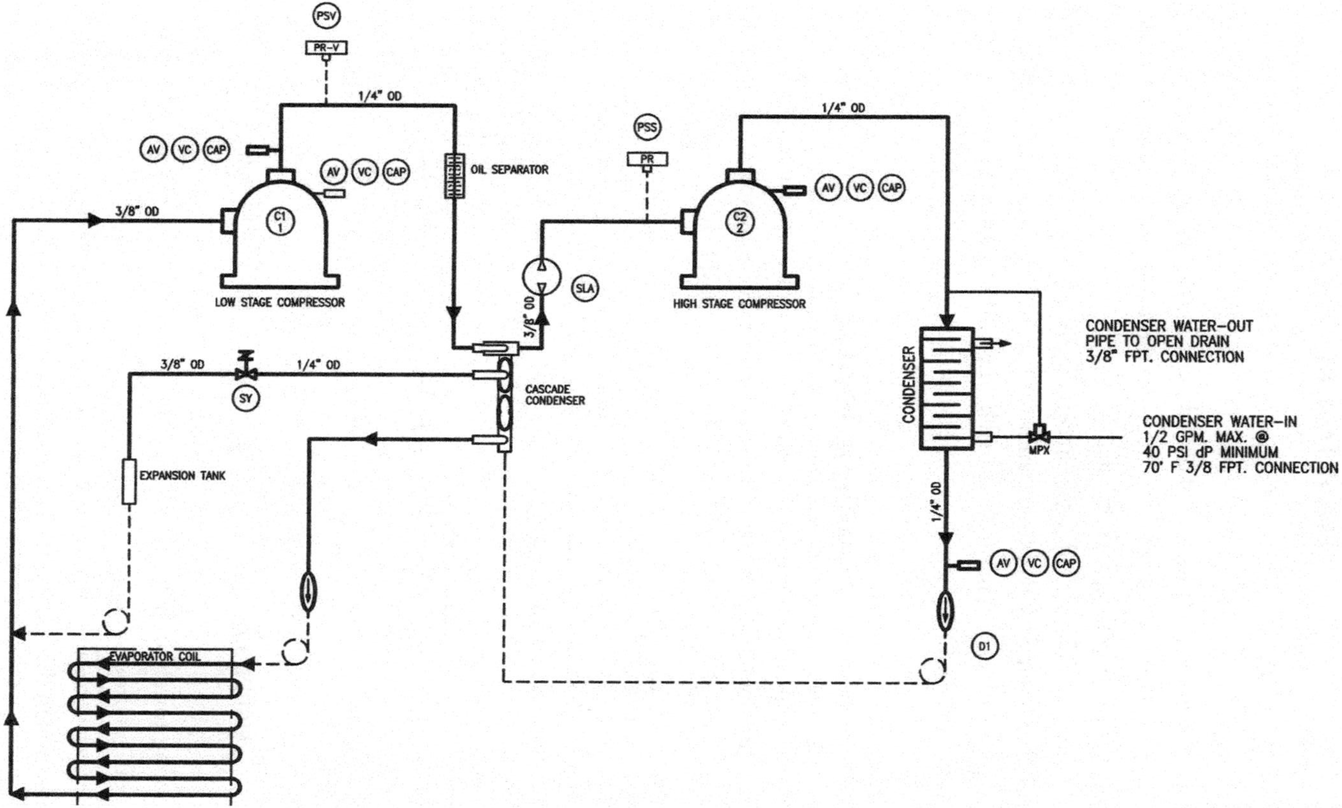

FIGURE 3.6 Cascade refrigeration system. (*Courtesy:* TPS, LLC)

Heat Pumps

Heat Pumps are DX systems, but the evaporator and condenser can be reversed so that they can switch between being used for heating and cooling. Heat pumps can use geothermal sources instead of outside air. Pipes are buried underground or in ponds and wells where these moderate temperatures are then used as a thermal source or sink for operating more efficiently.

Heat pumps can be a good choice in certain climates. Their cooling efficiency is measured by their SEER, which stands for seasonal energy efficiency ratio. A heat pump's heating efficiency is measured by the HSPF (heating seasonal performance factor) and COP (coefficient of performance).

Cooling Towers

Remembering our simplified chiller schematic, let's look at the cooling tower part of the cooling diagram, as shown in Figure 3.7. The cooling tower takes all the heat from cooling the building and sends it to the atmosphere.

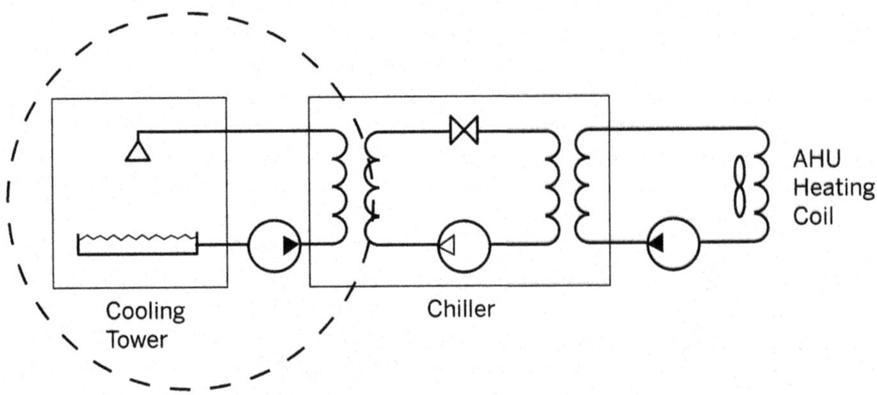

FIGURE 3.7 Cooling tower section of the cooling schematic.

Cooling towers usually rely on evaporation to get rid of the heat produced by the refrigeration system. Evaporation is a function of surface area, the cooling tower water-to-air temperature difference, and the humidity of the outside air blowing through the cooling tower. This is called *evaporative cooling*, and it is very efficient. It operates just like human sweat. The high-energy molecules go from the liquid phase to the gas phase and leave the rest of the water colder. Where the outside air is favorable, especially low humidity, this removal of latent heat is very efficient. The cooling tower efficiency is improved by using fans and fill material that expands the water out so that it has broader exposure to the outside air. Figure 3.8 shows a small cooling tower with buried water lines arising from the chiller condenser. Four large fans blow air upward through the fill material.

FIGURE 3.8 Cooling tower showing insulated lines going in and out.

DESIGN APPROACHES FOR COOLING TOWERS

There are two general categories of cooling towers, ***natural draft*** and ***mechanical draft***. Natural draft cooling towers are the things people confuse with nuclear reactors. They are the big hourglass-shaped concrete structures you see next to some very large power plants. Fans are used to move air in mechanical draft systems. This is done with either an induced draft or forced draft fan. Induced draft fans suck air out of the cooling tower, while forced draft fans push air into the cooling tower.

Crossflow designs move the air horizontally while the water flows downward and perpendicular to the air. Conversely, a ***counterflow*** design moves the air vertically upward, opposite to the downward-flowing water. Usually the water flows or is sprayed onto a fill material. The fill increases the surface area of the water for evaporation. Fill material is usually thin and weak and encounters all sorts of rough conditions, such as algae and bacteria growth, insects, fouling with debris, scaling from mineral deposits, and physical deterioration. This is on top of all the other challenges nature throws at outdoor equipment. Figure 3.9 shows both crossflow and counterflow cooling towers.

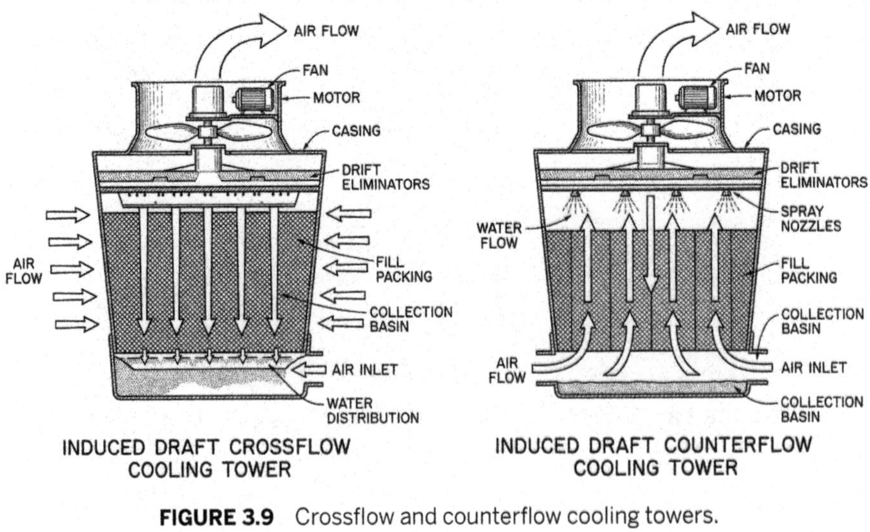

FIGURE 3.9 Crossflow and counterflow cooling towers.

Most cooling towers use an open-circuit system in which the water and air are in direct contact as described above. In the open-circuit design, the water system is hydraulically disconnected so the intake at the bottom of the cooling tower is gravity fed, not directly fed as would be the case with a closed-loop system. Closed-loop systems use finned coils and do not rely on water evaporation. This approach is less thermally efficient because it doesn't take advantage of the latent energy removal produced by evaporation. However, closed-loop systems save water, which is important in some areas. Some designs use gray water for this cooling.

Other forms of cooling the condenser side of the refrigeration system are common. These include straightforward air-cooled chillers that use fans to pass air over heat exchanger coils. These coils can be inside the chiller or remotely mounted outside.

Refrigerants

Talking about refrigerants can be tough because they are invisible, unfamiliar gases governed by laws and equipment manufacturers' requirements. And they seem to change a lot.

We want them to be safe, meaning nontoxic and nonflammable. We don't want them to hurt equipment, so they need to be noncorrosive. Finally, we want them to be kind to the environment, because there is a lot more leakage of refrigerants than anyone wants to admit. We would love it if nitrogen were a good refrigerant—it's the gas all HVAC technicians are most comfortable with. It's like playing with a good dog on a rainy day.

Many refrigerants are blends of chemicals that strike a balance between safety, noncorrosivity, and low environmental impact. Some, such as R-454B, are "mildly flammable," which is worrisome for those of you that are lighting up a torch every day and store refrigerants in your work vehicles. However, working with these refrigerants is what makes you a trained professional, so you are expected to understand their behavior, especially how it affects you and the public.

Knowing whether you are dealing with Class 1, 2L, 2, or 3 is important in understanding the flammability of refrigerants. Health impact (toxicity) is also a big concern when using refrigerants. Our biggest worry is that we are dealing with something that could hurt us without warning. If the classification starts with the letter B, watch out. It's toxic. If it starts with the letter A, it has low toxicity. As with mold, toxicity is a function of the amount of exposure. A little bit is not bad; a lot is dangerous. As the EPA likes to say (literally), "*Dosis sola facit venenum.*" Fortunately, the EPA translates it for those of us, like me, who struggled with Latin in high school. All I really remember was our lament: "Latin is dead, dead as can be. First it killed the Romans. Now it's killing me." Anyway, enough about my high school . . . The Latin term translates to mean "The dose makes the poison." That is, most substances are toxic in high concentrations and nontoxic in low concentrations.

Chemicals that are dangerous to humans with low concentrations and short exposure time are considered highly toxic. So refrigerants classified as A1 are nice and don't burn, while those classified as B3 are dangerously toxic and burn easily. The R-454B refrigerant mentioned previously is classified as A2L. The "2L" means it is a little less flammable than a "2." Check out ASHRAE Standard 34 for more information on this. Note that R-454B is near-azeotropic. I always thought this was a funny way of saying, "zeotropic, but not too much . . ."

You have a right to see an SDS (Safety Data Sheet) for anything you work with. This document replaces what us old-timers knew as an MSDS (Material Safety Data Sheet). There are a lot of details in these documents, but they are understandable. They all look kind of scary, so check out how frightening the SDS is for unleaded gasoline sometime to establish a framework to understand these documents.

WONDERFUL WATER

Water is an incredible molecule!

In addition to water's high specific heat that lets us store a lot of energy with a small change of temperature difference, water is unique because of its shape. The way the hydrogen and oxygen are attached is not in a straight line, but they are at an angle of 105°. This angle makes water molecules have a positive and negative end. The positive end is the side where the hydrogen molecules are oriented. These opposite charges allow water molecules to link together easily through what is called *hydrogen bonding*. This polarity makes water seem like a much bigger molecule than it actually is, and therefore it remains a liquid at room temperature. When not stuck to each other in a water molecule, oxygen and hydrogen link up with themselves (O_2 and H_2) but are still almost spherical and therefore don't easily stick to each other. Consequently, oxygen and hydrogen stay in a gas form unless the temperature is very cold. In fact, hydrogen freezes at −434 °F, which is almost at absolute zero. We are fortunate that water is in liquid form at earth's common temperatures—remember what our body is mostly made of?

Water's freezing behavior is also eccentric. Unlike most materials, it gets less dense as it gets close to its freezing point. It is densest at 40 °F rather than at its freezing point of 32 °F. What this means in practice is that as the surface of a water tank chills down to 40 °F, the 40 °F layer will sink and be replaced by the warmer water below. This movement continues until the whole volume of water is 40 °F. Only then will the surface chill below 40 °F. When water gets to its freezing temperature, it develops an ice layer that gets thicker but keeps the bottom part of the tank liquid for a long time. This is obviously a big advantage for marine life in ponds and lakes. The underwater creatures can hunker below the ice surface and survive the winter. Of course, a tank can freeze solid, and then the temperature can really plunge. It's this density reversal that makes water expand (therefore getting less dense) as it freezes and causes it to break stuff and causes us worries on cold nights. The surface forces that hold water molecules together are fairly weak. If the molecules get hot enough, the movement is enough to break these bonds. This thermal agitation is what turns ice into liquid water and finally into steam as shown in Figure 3.10.

Another unique characteristic of water is that it melts under pressure. This is why you can make snowballs. When you squeeze the snow, the ice particles melt under the pressure. When you stop squeezing and release the pressure, the liquid water instantly refreezes and sticks to itself. This pressure melting phenomenon is also why ice skates work so well—the focused pressure of the ice skate blade turns the ice into liquid water that provides a lubricated pathway.

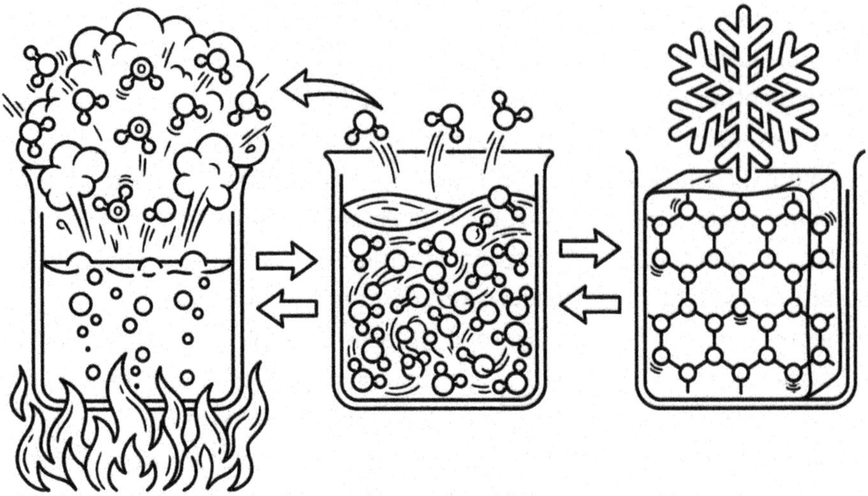

FIGURE 3.10 The surface forces of the water molecules get overwhelmed when they are heated to boiling and underwhelmed when freezing.

SAFETY SUMMARY
Refrigerants

The biggest issue with refrigerants is that they are asphyxiants, meaning they can displace air and can cause suffocation. There can be a scenario whereby the heavier-than-air refrigerant pools on the floor, and if the refrigerant causes you to faint, you fall into an even denser cloud of refrigerant. That scenario won't end well. Of course, refrigerants can be toxic and flammable, and they can produce frostbite. They should only be used by certified professionals. If a refrigerant charge is dumped from a tank or system, the process generally is to ventilate and evacuate. I've only had this happen once, and that was in a classroom setting and the students evacuated without any issues.

Unfortunately, there is a lot of leakage in large systems. One extreme case of leakage was when I was inspecting a failed chiller for an insurance claim. It turns out that the refrigerant line had broken (the evidence pointed to it having rolled) and all the refrigerant was discharged. The manufacturer shipped the chiller out without a speck of refrigerant.

Thermodynamics

I made an offhand remark about the second law of thermodynamics, saying that energy moves from hot to cold. This is true. Randomness always increases, and this requirement tells you which way a chemical reaction will go and indicates the tendency of molecules to go from being aligned in a liquid to unaligned highly randomized in a gas. However, the first law of thermodynamics is more relevant to studying refrigeration. The first law of thermodynamics states that energy is not created or destroyed but can change its form. This relationship is most commonly seen in the conversion of *potential energy* (such as pressure) to *kinetic energy* (velocity). The relationship shows how potential (and kinetic) energy can be converted to *temperature* to *work* (such as pumps and turbines). All these energy forms can be converted into each other. Yes, matter can be turned into energy, but let's not go nuclear at this point.

You can think of the first law of thermodynamics in its simplest form, namely conversion of pressure to velocity. If you fill up a balloon with air, it has a certain pressure. If you release the end of the balloon, that pressure is converted to velocity as the balloon flies through the air. *Surge pressure* (or *water hammer* in the case of water) is the opposite. If water is flowing along peacefully and you quickly close a valve, the water smacks into the valve and converts its velocity to pressure.

In HVAC, we think of a pump turning electric power into mechanical pressure and velocity. The pump does work on the system, and the response of the system is to increase the pressure and velocity of the fluid. If work is not added to a system, such as a pump adding or a turbine taking away, then velocity and pressure work against each other (sorry for the long sentence—you might want to read it a couple of times).

Changing velocity into pressure can be seen in surge pressure and static regain in ducts. Conversely, turning pressure into velocity can be seen every time you turn on a water faucet. Pressure and velocity are the key parts of the *Bernoulli principle*, where you get suction if you accelerate a fluid to a higher speed. This can draw ductwork inward and pull fuel into a carburetor, and I think it has something to do with airplane wings.

In the case of the cooling system, the refrigerant's phase change produces a change in potential energy, the compressor adds work energy, and the temperature changes in response; but it all has to balance out according to the first law of thermodynamics. Look into the *Joule-Thomson coefficient* to get good insight into how this works at a molecular level where surface forces fight thermal vibration to determine whether something will be a solid, liquid, or gas.

Absorption Chillers

The absorption refrigeration cycle is very good if you have cheap heat available such as from some kind of industrial process. Absorption chillers use water as a refrigerant. Isn't that nice! The refrigerant water is boiled in an evaporator that absorbs heat. The water vapor produced in the evaporator flows to a chamber that contains a water-absorbing hydrophilic chemical such as lithium bromide. You may see the term *hygroscopic* used, which is a category of hydrophilic material, but I like to use the term *hydrophilic* (because I can remember *Phila*delphia is the city of brotherly love). I can also contrast it with the water-hating materials called *hydrophobic* (we love our *phobias*). Our lives are hard enough without struggling too much with these terms.

The hydrophilic chemical, lithium bromide, absorbs the water vapor and then becomes a liquid, which is much like the condensation in the vapor compression cycle. The solution becomes diluted with water, and the water has to be removed in a concentrator chamber that uses heat to boil off the water. As you can see, heat is the energy source for this process. So if heat can be obtained cheaply, this is a great approach—always much nicer to use than chemicals.

Evaporative cooling will be discussed in more depth later. This method of cooling is usually the cheapest approach and is a very important way to get rid of the heat energy used in cooling towers and small systems working in dry environments.

DESIGN APPROACHES FOR PUMPS, COMPRESSORS, AND FANS

Because we are talking about compressing refrigerants and moving them around, let's introduce the category of pumps, compressors, and fans. This will be expanded upon later when we consider how to move energy around a building. In terms of pushing fluids around, there's a lot of ways to do it; we typically use fans and compressors for gases and pumps for liquids. A note on terminology: If a pump moves a gas, it is usually called a *fan*, *blower*, or *compressor*. As mentioned previously, a pump is any mechanical device that adds energy to a fluid.

Pumps

Centrifugal pumps are common in HVAC because they efficiently move high volumes of fluid. Because often the role in the HVAC is to move energy around, moving high- and low-temperature fluids is a very efficient way to move it. A centrifugal pump works by increasing the velocity of whatever it is pumping and throwing it at an outer casing. The geometry of the outer casing (volute) reduces the velocity of the fluid and converts it to pressure.

Figure 3.11 shows a centrifugal pump, a real workhorse in the HVAC industry. Centrifugal pumps normally work with the fluid being drawn into a motor-driven impeller. The impeller discharges the liquid into the outer casing, which has a distinctive spiral shape that gradually slows the fluid and therefore raises its pressure in accordance with the first law of thermodynamics that, in this case, says that potential energy (pressure) must equal the kinetic energy (velocity).

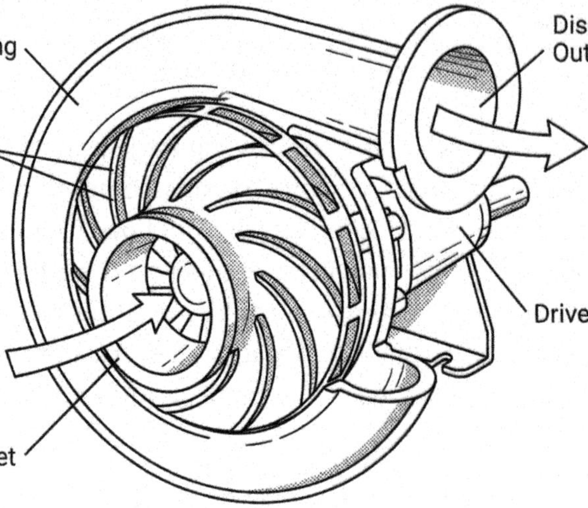

FIGURE 3.11 Centrifugal pump.

In liquid applications, *cavitation* can be a problem when the pressure produced around the impeller is reduced below the boiling point. Not only do these pockets of gas in the fluid around the impeller mess up the flow, but when they recondense and collapse, they can damage the surface of the impeller. The sound of cavitation is distinctively loud. The pump inlet pressure required to prevent cavitation is provided by the pump manufacturers as the *net positive suction head* (NPSH). Centrifugal pumps love to have a flooded suction where the fluid naturally flows into the inlet. Positive pressure at the inlet is even better. However, many pumps are capable of sucking fluids up to a certain height. Impellers can also be sequenced so that the discharge of the first impeller feeds into the inlet and another impeller. This series of impellers can be lined up to give a high-pressure output.

Positive displacement pumps are the most suitable design category if you need to move a small amount of fluid at a high pressure or if you want to produce strong negative pressure to draw liquid against gravity. These pumps move fluid around, just like what happens in your car engine where a piston draws air in and mechanically forces air out. Your heart and lungs are also positive displacement pumps. Positive displacement pump designs include diaphragm, piston, screw, and lobe pumps. Diaphragm pumps are commonly used in industry. They work by moving a flexible membrane back and forth, just like your diaphragm pushes and pulls your lungs. Twin-acting diaphragm pumps are common (Figure 3.12). A single shaft drives two shuttling membranes back and forth. This action reduces the fluid pulsations. There are many more varieties. One of the weirdest ones I have used was a progressive cavity pump. I used this to pump a very viscous slurry I was working with. It can also pump fish heads.

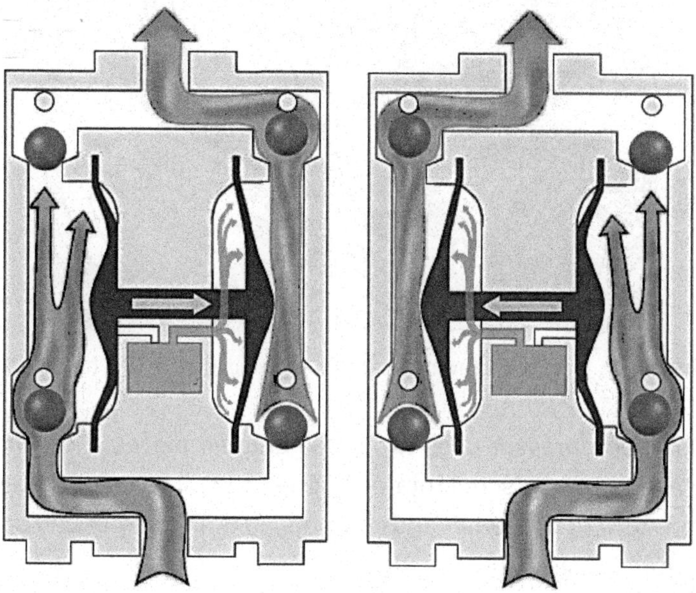

FIGURE 3.12 Twin-acting diaphragm pumps are commonly used positive displacement pumps. (*Source:* By Delange.mobi—Own work, CC BY-SA 3.0, Wikimedia)

Compressors

Centrifugal designs are common in large refrigerant applications like chillers. Other types of compressors commonly used in refrigeration systems are *screw, scroll, rotary,* and *reciprocating* designs. Like centrifugal compressors, screw compressors are used in large refrigerant systems. The screw compressor uses two interlocking helical screws (rotors) that look like worm gears. The helical rotors mesh together, with the space getting smaller and smaller as the refrigerant gets squeezed and compressed between the screws (Figure 3.13).

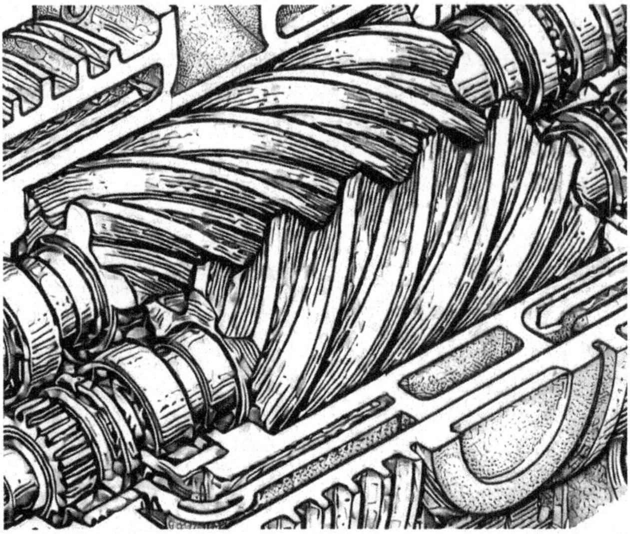

FIGURE 3.13 Detail of screw compressor showing interlocking helical screws.

A scroll compressor uses two scroll-shaped pieces. The inner scroll is fixed, and the outer scroll rotates (Figure 3.14). The scroll squeezes the refrigerant toward the center and compresses it. The scrolls can be pushed apart by liquid slugs without damage.

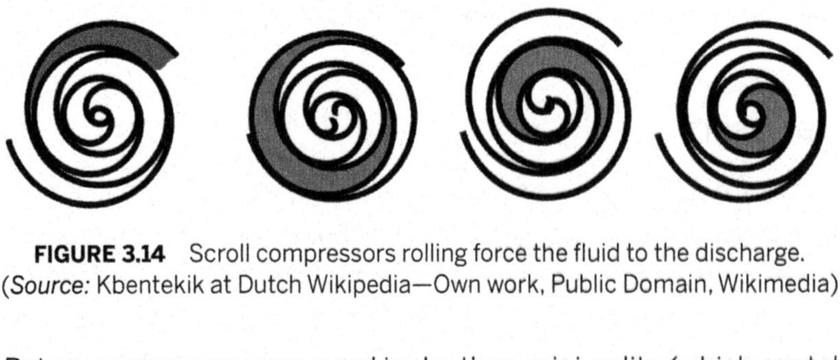

FIGURE 3.14 Scroll compressors rolling force the fluid to the discharge. (*Source:* Kbentekik at Dutch Wikipedia—Own work, Public Domain, Wikimedia)

Rotary compressors are used in ductless mini-splits (which are taking over the world—I have watched them replace ceiling fans in many parts of the developing world in the last couple of decades). They are also used in PTACs and mobile applications. Rotary compressors use eccentric rotating vanes or rollers inside a cylinder to squeeze the refrigerant in each chamber. The trapped gas is discharged when the volume is at its smallest and the pressure is at its highest.

Traditional reciprocating compressors are still around. They use a piston that slides up and down in a cylinder, working just like the pistons in your car engine. The vacuum produced by the downward movement of the piston draws in refrigerant. The refrigerant is compressed and is pushed out the discharge.

Fans

Fans are cheap, quiet, and reliable, and they move the end product of many HVAC systems, namely conditioned air. Most of us were introduced to fans when we were children and enjoyed the relief of a fan blowing across our sweaty body. There are parts of the world where fans are the only cooling available, and people move paperweights around to keep their fan from flinging documents. The modern fan was invented in 1882, which allowed those of us without palm leaf attendants to be cool also. These common fans, including ceiling fans, are *propeller axial* fans in which the blades spin around a central motor. The air moves parallel to the fan's axis. Other types of axial fans are the *tube axial* fan, which has a propeller inside a cylindrical housing, and a *vane axial* fan, which uses stationary straightening guide vanes at the discharge to reduce swirling of air and produces a more uniform outlet flow.

Centrifugal fans behave like centrifugal compressors where the velocity of the air increases in the fan and that velocity is then converted to pressure at the discharge. This category of centrifugal fans includes forward-curved, backward-curved, and airfoil designs. While these fans are included in most of the equipment you buy, their selection is a balance of efficiency, noise, and ability to handle dirty air. Forward-curved centrifugal fans curve the blade in the direction of travel. Backward-curved centrifugal fans have the blades curve away from rotation. Airflow centrifugal fans have blades shaped like airplane wings. While this design is very efficient, it can easily be degraded by dust and debris. The performance of these different designs is presented in data sheets. Figure 3.15 shows an electric motor driven axial fan and a squirrel cage fan. Squirrel cage fans are a common type of centrifugal fan.

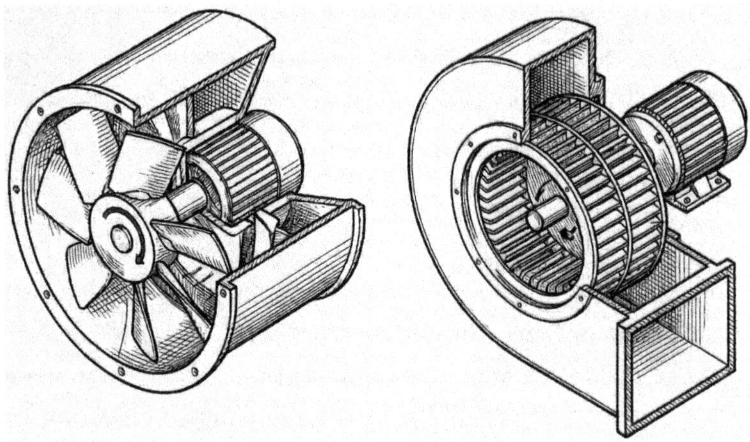

FIGURE 3.15 Axial and centrifugal fans.

There are some unique fans that are involved with specialty applications. For example, *tangential* (or *crossflow*) fans are used in air curtains. These are typically long fans that discharge air tangential to the impeller blades over their entire length. *Roof extractor* fans sit on top of a roof and draw air out from the space below. They can be powered or be wind-driven. *Bifurcated* fans are inline axial fans that are unique because the motor is separated from the fan so that the motor is not exposed to nasty stuff, like corrosive fumes and grease, that the fan might have to handle.

Fans have been designed for all kinds of specialty applications. Generally, your equipment has the best fan for its application. One of the bigger issues is the adhesion of dust and debris on fan blades. Another issue is that the large drive motors used on big air handlers use a lot of electric power and big motors have big inrush currents and restrictions on the number of starts per hour. We will look at this more when we talk about motors.

WHAT IS A FLUID?

The definition of a *fluid* is typically a liquid or gas. The specific definition is a little bit fussier because solids can act like liquids, and gases can act like solids. In practice we divide the world into solids, liquids, and gases (yes, there is a fourth state of matter called *plasma*, but that's not really an issue for us). We lump liquids and gases into the fluids category because both phases can't internally resist moving when a force is applied to them. When fluids have an applied load, the load produces a *shearing stress*, which is the sliding of molecules against each other. Solids, on the other hand, can resist these stresses. Simply put, this means if you push on a fluid, it will change shape, whereas a solid won't. All this is observed by watching the changing shape of a melting ice cube.

The resistance to flow is called *viscosity*. High viscosity means high resistance to flow. While heavy, viscous fluids resist applied forces, they cannot permanently resist them. Low-viscosity fluids, like air, can hardly resist shearing forces at all. However, there is an interesting exception to this behavior, and that is when gas velocities get to the speed of sound, or Mach 1. At this sonic boom point, the gas behaves more like a liquid. This liquid behavior is why explosives can kill people; it's like getting hit by a wall. (I guess. I've only gotten my clothes shaken by an acetylene explosion.) One way to think about fluids is to consider solids having an infinite viscosity. However, some solids flow over time. This behavior is called *viscoelasticity* which is the term used to describe solid-like materials that deform over time when a load is applied to them continuously.

SUMMARY OF WORRIES

Here are some things you should always be worrying about regarding cooling systems:

1. Refrigerants can suffocate people. If they aren't doing that, they can be flammable and toxic.
2. Superheat is important; liquid in vapor lines can cause damage. TXVs/EEVs are not magic wands.
3. Nature is trying to destroy your cooling tower and outdoor equipment. From lightning to bugs, your equipment is under attack. The list of methods is very long. Inside a building, your list is shorter. It can start with mice eating your electrical insulation; then you can move to piles of decaying rodents in unfortunate places.

FOUR

MOVING HEAT AND COLD

Piping and Heat Exchangers

Now let's consider the piping system that moves the hot or chilled water to the air handler. The piping for radiant systems is covered by these same principles. An interesting way to move heat and cold around is with district systems. District heating and cooling systems are seen in some large assemblages of buildings such as military facilities and universities. Using central systems can be very efficient, but they're fairly uncommon in the day-to-day work for an HVAC technician.

This is a deep engineering dive, so let's get specific about terms: Both liquids and gases are considered fluids; see the What Is a Fluid? sidebar in Chapter 3. When a gas flows at a high *Mach* number that is close to the speed of sound, it acts like a liquid and becomes incompressible. But for this discussion, I'll talk about liquids, although you need to recognize that gases operate in the same way, and we'll look at that when we consider the air distribution system after the air handler.

Drag

When heated or chilled water is pumped through a piping system, it encounters resistance to flow (drag) from two things. One is the friction in the pipe, and the second is from fittings and other geometry changes. While the friction between the water and the pipe causes drag, this is usually a small part of the total drag in the system. The biggest factor is usually fittings. These geometry changes produce *pressure drag*. Pressure drag (or form drag) is created by such things as tees, elbows, valves, and that sort of thing. Of course, squeezing liquid through a heat exchanger produces a lot of drag too, but I'm including that with geometry changes.

Neither water nor air wants to change direction quickly. This is why fish and airplanes are pointy at both ends. These shapes let the water and air gradually change direction. If you have abrupt changes in shape, you create a high pressure in the front where the air or water smacks into the shape. In the back, you get low-pressure eddies. A low pressure produced behind something moving through air, like your car, is like adding a giant vacuum cleaner behind the car. It wants to suck the car backward. You can see low-pressure eddies form behind bridge pylons in fast-moving rivers. You will also notice this effect if you hold your hand perpendicular to a stream of water.

You can remember the idea of pressure drag versus friction drag by sticking your hand out of your car window while driving. If your hand is parallel to the ground, there is very little drag. However, if you turn your hand at a right angle so the wind hits your palm, there's tremendous drag. This is true even though the roughness of your hand and the amount of area of your hand is unchanged in either orientation.

Figure 4.1 illustrates the differences between these two drags. Of course, they add together to create the total drag. The drag and boundary layer issues are important for sizing pipes and pumps as well as maximizing the effectiveness of heat exchangers.

On the math side, calculating the losses in straight sections of pipe and fittings is very easy. You can use the Hazen-Williams equation (super easy and given in the Appendix) or the Darcy-Weisbach equation along with the Moody diagram (messy but tells a great story). We are not looking at sizing equipment here, but it is surprisingly easy.

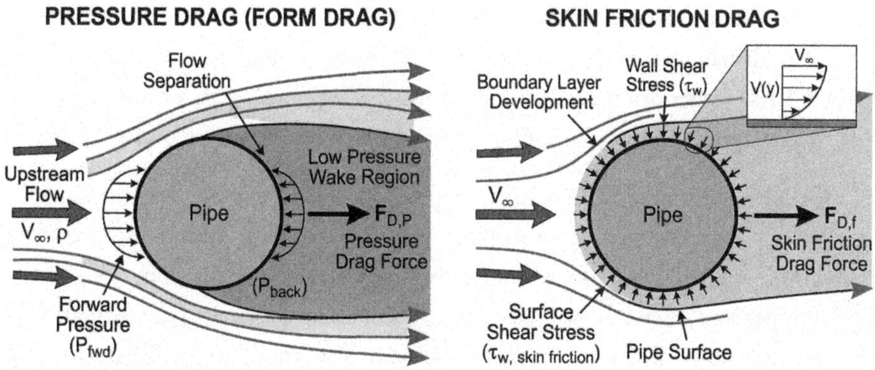

FIGURE 4.1 Pressure drag versus skin friction drag on a round pipe.

Piping Efficiency

The faster that water is pumped, the more drag it produces. Therefore, there's a balance between getting the most efficient pumping speed and the economics of pipe sizing. If you pump water at a very slow rate and at a low pressure, the pumping costs are low, but the pipe has to be very large. Large pipes mean a high installed cost. If you pump at high speed and high pressure, you can use smaller piping, but you have more electrical costs associated with pumping. This is why HVAC design usually starts with a recommended velocity.

Recommended velocities are based on experience, and they cover basically everything, such as:

- **Stack velocity.** Ensures the exhaust has enough inertia to clear the stack or exhaust flue.
- **Pump inlet and outlet velocity.** Avoids cavitation and maximizes pump efficiency.
- **Pipe velocity.** As discussed previously, this value balances pipe costs and pumping energy.
- **Air handler coil face velocity.** Prevents carryover of condensed water into the ducts during cooling and optimizes heat exchanger efficiency.

- **Neck velocity and throw velocity.** Provide comfortable and effective air distribution in a room so it's not too drafty or noisy for the occupants. These values drive the throw and spread of supply air. This air distribution concern also tries to minimize stagnant and stratified air in occupied zones.

Other recommended velocities are provided for things like fuel and instrumentation. These are usually important starting points for design and are often based more on real-world experience rather than abstract rules of physics.

In pipes and ducts, the velocity lets you size them properly. The calculation is simply area equals flow rate divided by velocity. You know the required flow rate based on thermal load calculations, and you have a recommended pipe velocity. Divide the two (watch units!), and you have your pipe or duct area. Area is equal to the inside dimensions of the pipe or duct.

Fluid Dynamics

Let's look deeper into how fluids flow, called *fluid dynamics*. This study of fluids in motion is important in both pipe and duct design. While this chapter is focused on moving heat and cold through pipes, all of this presentation on fluid dynamics applies to fuels and air also.

Fluid dynamics is rooted in the notion that mass is conserved—what goes in must come out. It is also grounded in the theory that energy is conserved, which comes from the first law of thermodynamics we talked about before. Because fluids are moving, the changes in direction and conversion between velocity and pressure are part of this energy balance. The other part of this discipline is friction or drag, which is what we will consider here.

Laminar and Turbulent Flow

There are two types of flow that are encountered. You've heard the terms before: one is called *laminar* flow, and the other is *turbulent* flow. The speed of the flow determines whether something is laminar or turbulent. Laminar flow tends to be self-correcting. If it is disturbed, it quickly returns to its former condition, while turbulent flow just goes crazy and moves around chaotically. These two types of flow produce different amounts of drag. You can tell what type of flow you have by knowing its velocity. More specifically, you need to know its *Reynolds number*, which tells you whether the flow is likely to be laminar or turbulent.

The Reynolds number measures the ratio of velocity over viscosity; more specifically, the Reynolds number presents the ratio of inertial force (which includes mass) to the viscous force. Therefore, fast-moving fluids have a high Reynolds number, and slow-moving fluids have a low Reynolds number. Because the viscosity is in the denominator (I remember this as the *d* in *denominator* stands for "down," as in the down side of the fraction; *denominator* is so wordy), a high-viscosity liquid, like No. 6 oil, will have a low Reynolds number. On the other hand, low-viscosity fluids, like air, will have a high Reynolds number. Again, high-velocity fluids have a high Reynolds number; low-velocity fluids have a low Reynolds number. Basically, Reynolds number is a fancy way of saying *velocity*, and it's used in a geeky way that I'll talk about later when we look at heat exchangers.

The Reynolds number has no dimensions, so its meaning applies to all fluids. Generally, liquids with a Reynolds number (abbreviated *Re*) above 2,000 are turbulent, and a Reynolds number below 2,000 is laminar. This is one of many dimensionless parameters used in fluid dynamics. The most common is the *Mach number* (Ma), which is used to give the ratio of the speed of something compared with the speed of sound. As noted earlier, Mach 1 equals the speed of sound. This value of Ma = 1 gives a sonic boom, and the air acts more like a liquid.

Boundary Layer

Laminar flow and turbulent flow have very different flow characteristics and heat transfer characteristics. Therefore, both fluid dynamics and pipe design theory mix with heat exchanger theory at this point.

The reason both heat transfer and pressure drops are affected by laminar (low Re) and turbulent (high Re) is because the thickness of a thin film called the *boundary layer* changes between these two types of flow. The boundary layer is a skinny layer of slow-moving particles that drag against a pipe or duct. The slow movement is caused because the molecules closest to the pipe or duct wall basically get stopped by the surface. Because the molecules are "sticky" due to their viscous surface forces, the stopped molecules slow down the ones next to them. This continues outward until the molecules get past the boundary layer and into the fully developed flow area in the middle of the pipe or duct.

Laminar flow has thick boundary layers; turbulent flow has thin boundaries. Often you think turbulent flow is bad, but actually it is good for a variety of reasons. The thin, turbulent boundary layer is less insulating and allows better heat transfer. It also helps reduce pressure drag. A thin boundary layer also has better adhesion to a surface. This is why golf balls have dimples. The dimpled surface roughness "trips" a laminar flow into turbulent flow. By doing this, it allows the air to wrap around the golf ball. If the boundary layer separates from the back of the golf ball, a low-pressure void is produced, sort of like having a vacuum cleaner sucking air from behind the ball. This makes the pressure drag I mentioned earlier in this chapter usually the biggest contributor to pressure drop. Pressure drag is produced in fittings and especially heat exchangers. It is why you want gradual changes of geometry and velocity in both liquid pipe and air ducting. This is usually far more important than the *friction drag* produced by the roughness of pipes and ducts. But it's not that easy. Friction increases with speed. ("Thanks for telling me the obvious.")

While higher speed produces lower pressure drag, the higher velocity produces more friction drag. So there is a balance. However, if a fluid is almost turbulent, you can help it become turbulent by adding roughness. This doesn't increase the friction drag too much (because the speed is the same), but it greatly reduces the pressure drag (because the sticky boundary layer tends to reduce eddies).

So how do you remember all this stuff? First, think of a fast-moving, turbulent boundary layer as if it were a plastic cling wrap—it's very thin and sticks to surfaces. Second, think of the laminar boundary layer as if it were a foam mouse pad. It doesn't conform well to surfaces, and it's thick. More specifically, you can think of molecules, say of water, flowing through a pipe. Some are grabbing the pipe wall and are basically stopped. The molecules in the middle of the pipe are going fast. These molecules are connected to the slow molecules by surface forces. It's like they are holding hands. The viscous drag between the molecules is only a big issue near the pipe wall, where the fast-flowing molecules are being slowed down by the slowpokes in the boundary layer.

This boundary layer stuff affects heat exchange too. With turbulent flow's thin boundary layer, there are more high-velocity molecules flowing through a pipe. This improves the efficiency (properly called *effectiveness*) of heat exchangers. A turbulent boundary layer (along with a high temperature difference and a lot of area and flow rate) allows a high heat exchanger effectiveness. The problem with a high Reynolds number (that is, thin, turbulent boundary layers) is that it comes at the cost of high pumping energy. Heat exchangers are refined to give a turbulent flow at the lowest velocities to improve efficiency of the pumping system. In fact, high-efficiency heat exchangers are sometimes called *low Reynolds heat exchangers* (by snobs) because they get good heat transfer at a low flow rate. Remember, a low pumping flow rate means less electricity consumption. But if the pump rate is too slow, you don't get a turbulent boundary layer, which you want in a heat exchanger. These high-efficiency, low-Reynolds (let's be snobs together) heat exchangers are designed to trip boundary layers from laminar to turbulent layers. They often use mechanical features called *turbulators*—just like the dimples in a golf ball. Yes, turbulators increase the friction drag, but their improvement in heat transfer and/or reduction in pressure drag can make them useful. Figure 4.2 shows one type of turbulator, but they can take many forms. The design takes high-speed (or high-Reynolds if you are geeky) laminar flow and trips it into turbulent flow.

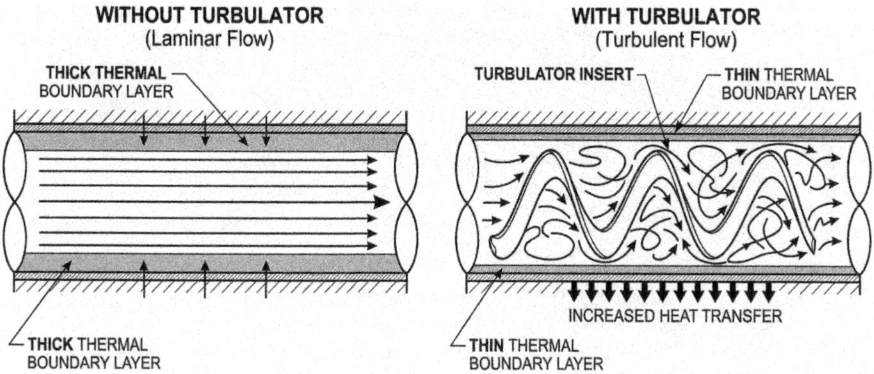

FIGURE 4.2 Turbulators can trip high-speed laminar flow into turbulent flow.

This is kind of challenging to understand, so here is a summary of Reynolds number, flow characteristics, and performance:

- Slow or thick fluid = low Reynolds = laminar (probably) = thick boundary layer = high pressure drag = low heat transfer
- Fast or thin fluid = high Reynolds = turbulent (probably) = thin boundary layer = low pressure drag = high heat transfer

Many types of heat exchangers are available, and they have been optimized for specific applications. There is a balance between the effectiveness of a heat exchanger and the amount of energy required in terms of pump and fan requirements. It is easy to get excellent, high-effectiveness heat exchange if you are willing to pay for it. The goal is often to take the amount of energy required to push the fluid through the heat exchanger and balance it with the cost of the heat exchanger itself. Finned-tube heat exchangers are common for water-to-air heat exchange in coils. Shell-and-tube and plate heat exchangers are common for fluid-to-fluid heat transfer. Figure 4.3 shows one variant of a shell-and-tube heat exchanger, and Figure 4.4 shows a plate heat exchanger. Some larger plate heat exchangers are built on a frame and can be taken apart and expanded.

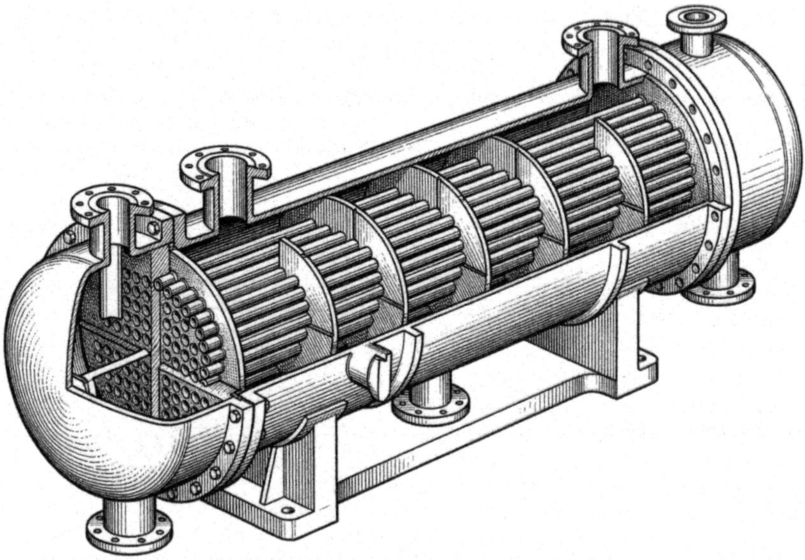

FIGURE 4.3 Shell-and-tube heat exchanger.

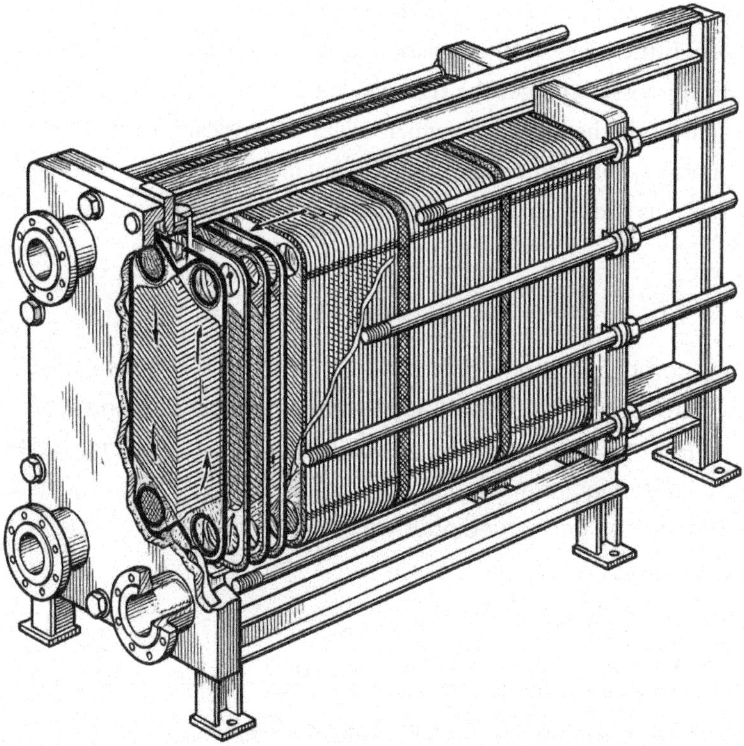

FIGURE 4.4 Plate heat exchanger.

Summary of Drag and Sizing

The two types of pressure drag that an HVAC system has to overcome are (1) friction and (2) pressure drag (produced by fittings, heat exchangers, filters, and other equipment). The starting point for engineering design of piping systems is a recommended velocity, and this is also true for duct systems (and almost everything else in the HVAC world). For a desired velocity, it's an easy calculation to figure out the pipe and duct sizing. Area is equal to flow divided by velocity. This area calculation gives you the required size for your pipe or duct. The flow comes from things like thermal load calculations, while the velocity is usually based on rules of thumb.

The basic engineering considerations are first to move the right amount of heat through the pipes or ducts (based on load calculations) and then strive to reduce fitting losses and get the right velocities through heat exchangers, coils, pumps, fans, and other devices in which velocity is critical to performance.

Economics, not engineering, drives the balance between the cost of operating pumps and fans and the sizing and installed cost of the pipes and ducts.

DESIGN APPROACHES FOR PIPING

Pipes are commonly made from metal or plastic. There are some outliers that aren't used in HVAC such as clay, concrete, and Orangeburg pipes. Clay pipe is very old. In its modern form, it is called ***vitrified clay pipe*** (VCF). Concrete is used for large drainage. Orangeburg pipe, which was made from wood pulp and tar, performed poorly over time and is now obsolete.

Plastic

Plastic pipe has revolutionized pipe design and installation. PEX, HDPE, PVC, CPVC, and ABS are the most common cascade of acronyms for modern piping. They are cheap, available in long coils, smooth walled, and noncorrosive. They are hard to beat in some applications. Fittings are easily installed and sometimes fascinating to work with (like exotic tapping tees and electrofusion HDPE). The biggest drawbacks, compared with metal, are the potential for smoking when heated, poor high and low temperature strength, and low mechanical strength.

Metal

Copper, steel, and cast iron fit in this category. Metal doesn't create smoke. While metals can be softened at high temperature, they are reliable materials in extreme environments. Material behavior will be discussed in detail in Chapter 5, "Materials in HVAC Systems." Although I won't discuss cast iron here, I will note that I have seen cast iron used in some municipal water applications, where there is resistance to HDPE. But the use of cast iron is dwindling.

Copper

Copper is a wonderful material; it's too bad it's so expensive. It does everything well. Soft ACR tubing is a delight to work with (ACR is an acronym for air conditioning/refrigeration). I enjoy watching the changing colors under heat and the metal swell under the hammer when making swaged connections. You learn a lot of things working with copper, such as how cheap flare blocks never make good flares and weird things that happen to joints when you apply too much heat.

Steel

Schedule pipes, like 40 and 80, are thick and robust and have a great track record. Thin-wall pipes are designed to handle a specific application efficiently. They are extremely common in fire protection applications. Thin-wall pipes have to be individually approved and are defined by ASTM as pipe that has a thickness that is 3% or less of the outside dimension. The attachment systems include techniques like grooved couplings and plain-end socket fittings. Schedule pipes are used for higher-pressure applications, and the attachments are usually threaded or welded.

Heat Transfer

There are five ways in which energy is moved around by heat exchange: (1) conduction, (2) convection, (3) radiation, (4) the creation (or depletion) of energy, and (5) the moving of heated (or cooled) materials. Of these, conduction, convection, and radiation are truly heat transfer and move only energy; the other two are different. Figure 4.5 shows a simple way to remember the three types of heat transfer. I don't know if sticking your finger into a flame is a good example of conduction, but you get the idea.

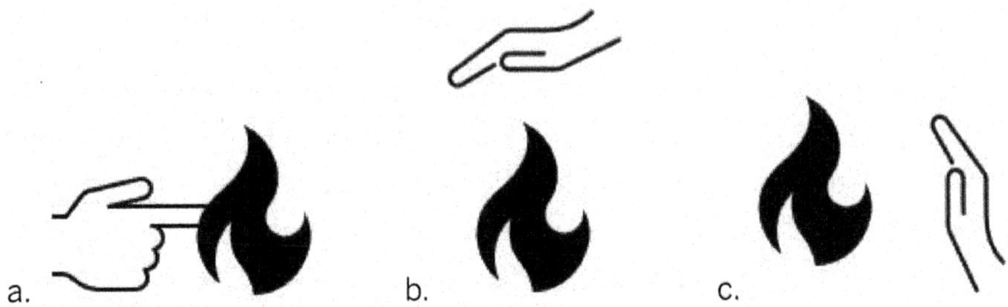

FIGURE 4.5 Conduction (a), convection (b), and radiation (c) from a flame.

Conductive heat transfer is material-to-material heat transfer, and it's a function of the contact area, temperature difference, and the material's thermal conductivity. *Convective* heat transfer is much the same as conduction in terms of the math, but instead of thermal conductivity, you look at the convection coefficient, which is largely related to the velocity of the fluid flow. *Radiation* is a function of temperature difference, exposed area, and emissivity (or absorption) of a material. This becomes an important issue when doing heat load calculations, which we're not discussing here, but they are discussed in the Appendix.

Conduction, convection, and radiation heat transfer are always a function of the area and temperature difference. That's the science. It doesn't change. The thermal conductivity affects conduction, the velocity affects convection, and the emissivity or absorptivity affects radiation.

One of the other considerations, *heat generation*, is something we talked about when we looked at combustion. As its name implies, heat generation is a transfer of heat from chemical energy to thermal energy.

Finally, moving material around, called *mass transport*, is like opening the door of your house on a cold day. It's moving matter from one place to another, so if you open the door and let cold air inside, those cold outside molecules will mix with the warm inside molecules in your house. This mixing causes the inside house temperature to get colder. You didn't move energy; you moved physical molecules. And so in that way it's different than the others. Mass transport is caused by pressure difference, not temperature difference. Pressure from wind or the stack effect will push air molecules around, and the pressure doesn't care about the molecule's temperature. If you open a window on a windy day, you will get hot or cold air blowing in. The wind didn't ask about the air temperature before it pushed the air into your house. This mass transport phenomenon is what you consider when looking at *infiltration* losses in heat calculations.

Thermal Storage

An important cousin to all this heat transfer stuff is *thermal storage*, which is vital for some architectural designs and HVAC systems. Heat (or coldness) can be stored in materials, and the amount of energy that can be stored is a simple mathematical relationship. The stored energy is equal to the mass, the specific heat of the material, and the temperature difference. *Specific heat* is an intrinsic property of a material. An important thing to remember is that the specific heat of water is over four times the specific heat of air. The units of specific heat are per pound, and air is much less dense than water; therefore it takes a lot of air to move the same energy as just a little bit of water. This highlights the efficiency of moving energy by water.

Summary of Heat Transfer

You can remember some of this stuff this way: Think of conduction as putting your finger in a candle flame, convection as putting your hand above a flame, and radiation as putting your hand to the side of the flame. Think of heat generation as the wax that is burning and turning into heat energy. Mass transport is just like when you left the door open on a cold winter day and all that cold air flows in—which may very well blow out the candle flame you keep burning yourself with.

Water Treatment

Water treatment is used in pipes and heat exchanger systems to reduce microbial growth, pipe corrosion, and scale buildup. The water is kept neutral with pH adjusters, and other chemicals are used to keep the heat exchangers clean, pipes efficient, and bugs from growing.

Corrosion is often combated by chemicals that react with oxygen, such as sodium sulfite (Na_2SO_3). Oxygen reacts with sodium sulfite to produce sulfate (SO_4^{2-}) The little (superscript) "2–" means that there is a negative charge to the molecule. This reaction grabs oxygen from the water, sometimes referred to as oxygen scavenging, which is the most common contributor to corrosion. Remember, corrosion can be produced by other means as described in the Corrosion section in Chapter 5.

Scaling from minerals, such as calcium and magnesium, is often done with filtration, zeolites, phosphates, and other chemical means that cause the minerals to stick to them or by actually filtering them through their molecular structure. This molecular structure is called, appropriately, a *molecular sieve* and can trap specific molecules. It is amazing how important the shape of molecules is for many things. You wouldn't guess that sometimes chemistry is like a jigsaw puzzle.

Biocides are important for eliminating disease-causing *pathogens*. Pathogens are any bacterium, virus, or fungus that causes harm. The bacterium *Legionella* is one of the most famous in the HVAC world. It causes lung inflammation. *Legionella* growth is especially a problem with misting devices in open systems like cooling towers. *Legionella* is usually transmitted to humans as an aerosol. A protective biofilm protects some bacteria, like *Legionella*. Biocides have been developed to kill *Legionella* by breaking down the protective biofilm and then attacking the bacteria by destroying cell walls and other internal cell structures. Algaecides are a type of biocide that destroys algae cell walls or stops their cell division.

Venting and Surge Pressure

Air is important in water systems. In drainage systems it's vital to allow air to enter. In HVAC piping systems, you want to get rid of it. Air dissolves in water, and when pressure is reduced, it wants to come out just like the carbon dioxide coming out when you open a soda bottle. Less pressure means less dissolved air. Water temperature also affects how much air can be absorbed. Colder water will dissolve more air than hot water. Fish and other marine creatures get more oxygen in cold water than hot water. Note that oxygen dissolves better than nitrogen in water. Controlling the presence of air in liquid systems is an important design requirement.

When water stops quickly, a huge surge pressure arises because it's incompressible, which can damage equipment and bang around pipes. I've seen pipe hangers swaying like a child's swing when this happens. Water hammer arrestors give a chance for the water to slow down more gradually.

SUMMARY OF WORRIES

Here are some things you should always be worrying about regarding piping systems:

1. Water expands as it freezes.
2. Materials expand and contract with temperature.
3. Fast-acting valves create a surge pressure.
4. When fluids leak out of a pipe and spray out, the fluid gets atomized, which gives it a lot of surface area. I mention this only because it is a safety issue with fuel. Water just makes a mess. Steam will kill you.
5. Velocity guidelines are the basics for most designs. If the velocities are wrong, systems don't work well.
6. Fluid viscosity can change in peculiar ways for some liquids. This is rare.
7. Sometimes pipes are used as columns, and they are vulnerable to buckling.

While thermal expansion of pipes and ducts is always a consideration, the other stuff is less of an issue, except freezing water, which you would know if you had sufficient interest in HVAC to enter the profession and read this far! We'll look at some of this in detail later.

SAFETY SUMMARY
Pressure Relief Valves

Generally, pressure relief valves are for protecting equipment, while pressure safety valves are a last resort to save catastrophic failure and protect people. Safety valves operate quickly and reliably.

Pressure safety and relief valves work with a spring pushing against a sealing disc (sometimes called a *poppet*) that keeps the valve discharge closed. If the pressure against the disc exceeds the spring force, the valve opens and the fluid is routed out of the valve and directly into the atmosphere and sometimes into a receiving tank. The initial pressure required to move the disc is appropriately called *cracking pressure*. For high-pressure, high-volume applications, pilot-operated relief valves are often used. This design uses a small pilot line to move the disc assembly and release pressure. Pilot-operated relief valves don't start opening until near their set point so they don't waste as much fluid.

Figure 4.6 shows a pressure relief valve in the closed position. Increasing pressure pushes the compression spring upward and thereby opens up the discharge port. While the design seems simple, it's vital that these valves never fail, so there is a lot of attention given to the myriads of details and manufacturing connected with these important devices.

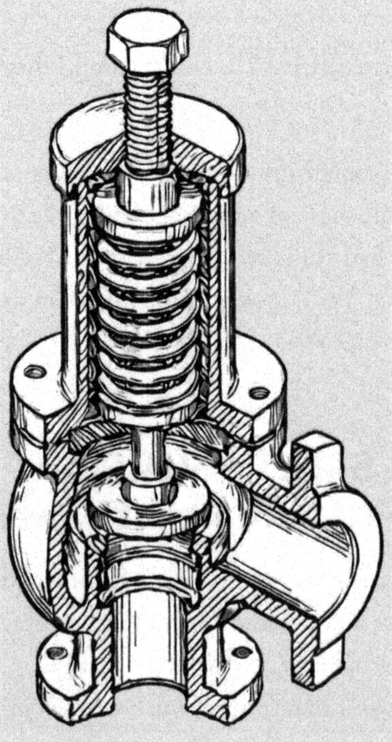

FIGURE 4.6 Pressure relief valve.

Pressure safety valves are a subset of pressure relief valves that are a fail-safe for pressurized systems. Related valves can include temperature trips also. This category of valve can also be called a *TPR valve*, *temperature and pressure valve*, or *pop-off valve*. ASME has a long, proud history of designing these things, motivated by the explosions of early boilers. These valves have saved many lives.

It's important that the line going from the piping or tank be directly routed to the pressure safety valve without any valves or other obstacles in between. The discharge of the valve must be routed away from human occupation. I remember being corrected on this one time when I had a pressure safety valve discharge located pretty much at eye level.

I did a forensic analysis of a large ash silo collapse where the 60-foot silo got crushed like a beer can. The vacuum pump that draws ash from the bottom of the boiler's hoppers was turned on while the intake was valved shut. The powerful vacuum pump caused the thin-wall silo to buckle inwardly. A simple "rupture disc," which will break open at a vacuum set point, should have been installed in the system. This would have saved the expensive silo's demise. In commercial projects, these types of control logic failures and design omissions occur. Not everything is idiot proof.

Pressure relief valves (and rupture discs) wait as sentinels of safety. We often push them into the background of our thinking, but they are worth a quick inspection to make sure there's nothing that would keep them from doing their important job. See ASME Section VIII for more information on standards for design. I am thankful so much talent is poured into the details of these things.

SAFETY SUMMARY
Surge Pressure

Fast-acting valves can cause a lot of movement in piping systems. I remember seeing one of my piping systems sway like a swing set when one of the valves shut closed. I was impressed to see the whole line of pipe move nearly a foot as it swayed under the pipe hangers. While this might not seem like a safety issue, having a mass of pipe swinging around can do terrible things. Look at how beam clamps work and you should be worried.

Hydraulic accumulators or surge tanks (or water hammer arrestors for small systems) can be used to handle surge pressure elegantly as opposed to exclusively using highly rigid piping. Surge relief valves are fast-acting valves that can reroute flow and reduce surge pressure. Slow-acting valves, like many large motorized valves, reduce this problem. There are other techniques too, such as using variable-speed drives to gradually slow down fluid flow or putting loops in the piping design. Air chambers/dead legs are not ideal because they retain stagnant water, which can harbor bacteria, plus the air in the chambers can get absorbed into the water and degrade the performance. ASME Codes B31.4 and 31.3 give guidelines for designing systems to handle surges and for setting limits on the amount of surge. Figure 4.7 shows a surge tank with the diaphragm that flexes to absorb all the spikes of pressure.

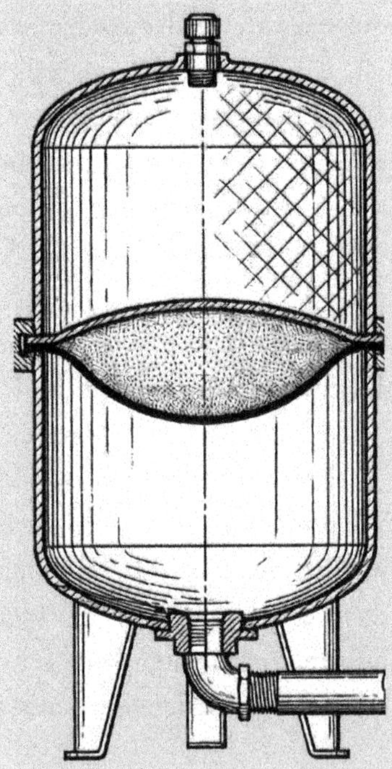

FIGURE 4.7 Surge tank.

SAFETY SUMMARY
Thermal Expansion

I encountered a long PVC piping system that I didn't trust. On a cold day, I took a hacksaw to it, and as I was about to finish the cut, the pipe snapped and pulled back nearly a foot in each direction. I'll never forget that incident. It was a dramatic illustration of the force of thermal expansion. You have probably seen the scraped paint on pipe supports. Thermal forces either produce a dimensional change or increase internal stress. Something will happen!

The coefficient of linear expansion provides a value for how many inches a material expands for a given length for every degree Fahrenheit it is heated. That is, a 120-inch-long steel pipe expands 10 times as much as a 12-inch-long steel pipe. These coefficients are provided in the Appendix and indicate the expansion for every inch of length. Basically, you need to know:

- Plastics expand over 10 times more than metals.
- Aluminum expands at about twice the rate of steel.
- Copper expands by about 1½ times the rate of steel.

There are many design approaches to adding sufficient flexibility in piping and ducting to handle thermal expansion. Don't lose sleep on frigid nights by worrying if you have enough expansion joints or flexibility.

RHEOLOGY

Some fluids act much differently than water, and I'm not talking about viscosity increasing at low temperature. I'm talking about viscosity changing with pumping speed, which is just weird. These fluids are called *non-Newtonian*, meaning they can behave in bizarre ways. This is rare in HVAC, but it does occur with some slurry fuels where solids and liquids are mixed.

Rheology considers how fluids behave when they have an applied stress—that is, when they are getting pushed around by pumps and nozzles. The rheology of non-Newtonian fluids is such that the viscosity becomes dependent on the pumping speed, or *shear rate* in engineering parlance, rather than just temperature. The viscosity can go up or down as the fluid is pumped fast or atomized.

You always want to know the rheological properties for the fluids you're dealing with, but fortunately water and air are Newtonian fluids and don't get weird on us; they have a linear relationship between the viscosity and the shear rate.

Summary of Fluid Dynamics and Heat Transfer

Let's try to squeeze a lot of information about the giant engineering disciplines of fluid dynamics and heat transfer into a summary.

The laws of fluid dynamics include the following:

- Friction drag and pressure drag produce the total drag in pipes and ducts.
- Reynolds number measures velocity (yes, it is actually inertia divided by viscosity), and this tells you whether you have laminar or turbulent flow.
- Turbulent flow has a thin boundary layer, so it wraps around shapes better, reducing pressure drag (and increases heat transfer). It's like plastic cling wrap.
- There are some weird concerns too, such as thermal expansion, freezing water, and changing viscosity.

Heat transfer occurs by a combination of these five methods:

- Conductive heat transfer
- Convective heat transfer
- Radiant heat transfer
- Heat creation
- Mass transport

FIVE

MATERIALS IN HVAC SYSTEMS

The materials used in HVAC systems vary widely. We use steel and iron for strength, and we use copper and aluminum for thermal conductivity and corrosion resistance. We use plastics to save money and reduce weight. We use foam for insulation and many specialty materials such as desiccants, seals, and adhesives. Materials have intrinsic properties that don't change when you make different shapes from them. Some of these properties include thermal conductivity, electrical conductivity, corrosion, strength, and strain. Strain is the amount a material stretches when a load is applied to it. Let's take a look at these properties. But first let's consider an overview of common materials you might work with.

Summary of Material Properties

Metals

Steel has the highest strength-to-weight-to-cost ratio available. You get a lot of mechanical strength for the money. Carbon steel is vulnerable to corrosion. Stainless steel, or more properly chromium steel, reduces the potential for corrosion, but it's expensive. Cast iron is also an iron and carbon-based material that provides a high strength for a low cost. But it is heavy. Copper and aluminum are wonderful materials because they provide high thermal and electrical conductivity and corrosion resistance. Like stainless steel, they develop an oxide, or passivated, layer that protects the underlying material from corrosion. Copper is the principal material in brasses and bronze alloys. Aluminum has a wide range of strengths and can give extraordinarily high strength-to-weight ratios.

Plastics

Plastics have been working their way into all elements of HVAC. The bad part of plastics is that they do not perform as well as metals for high- and low-temperature applications and can generate toxic smoke. However, we know the good part of plastics too. They are noncorrosive, cheap, and lightweight. In piping applications, they are easy to manipulate and connect and are available in long lengths. One behavior that is different in metals is that they tend to be viscoelastic, meaning they can gradually change shape over time when a load is applied to them continuously. Plastics can be vulnerable to ultraviolet light, and their fatigue and impact properties vary widely.

Table 5.1 provides thermal conductivity for common materials. Note that copper conducts heat 8 times faster than steel. Aluminum conducts heat at less than half the rate of copper. Plastics hardly conduct heat at all.

TABLE 5.1 Typical Thermal Conductivity for Common Materials

Material	Thermal Conductivity (BTU/(hr ft °F)
HDPE	0.5
PVC	0.2
Steel	30
Aluminum	95
Copper	225

Insulation

Insulation is commonly made from foam, which is usually a matrix of plastic and air. Because foams are mostly air, they are not very strong, ranging from weak packing foams made from polyurethane to stronger high-density foams used in furniture. However, foam is light and a good thermal insulator. Closed-cell foams have sealed air pockets and therefore don't absorb moisture. Insulating materials can be developed from combinations of materials such as fiberglass, aluminum, plastic, and paper.

Ceramics

We don't often deal directly with ceramics or refractory materials, but they are important in heating systems. Ceramics are almost the opposite of plastics. They own the high-temperature world but are brittle and hard to work with. Ceramics are an excellent material for certain applications, especially where they operate heroically on the combustion side of the HVAC by containing raging fires. Ceramics are strong, brittle, and chemically impervious materials that can operate at very high temperatures. They don't melt or corrode; they handle thermal shock and insulate. Insulating the fire keeps the combustion chamber hot, which is vital for combusting heavy oils and solid fuels.

Adhesives

Adhesives connect surfaces together, and they have saved the day in many cases. Not only do they repair things, but they have long been the default joining method for some materials such as PVC. Adhesives often rely on a solvent that evaporates and allows for tight surface bonding of materials. The surface bonding can either use a mechanical locking process, such as when the adhesive seeps into the pores of a surface, or use molecular surface forces to produce the adhesion. PVC cement is a solvent adhesive that softens the PVC and allows cross-links to be formed between the two mating PVC materials. Some adhesives, such as epoxies, do not rely on evaporation of a solvent but depend on the chemical reaction of a two-part system.

Other material considerations are interactions, such as fuel interacting with elastomeric seals or aluminum interacting with copper. A couple of other things to consider with materials are (1) they act like springs and (2) they expand and contract with temperature.

Material Characteristics

We use materials to contain energy carriers like water, fuel, and refrigerants. However, we rely on materials to enclose and support equipment. Therefore, we should take a look at how materials behave when stressed. While the focus on HVAC design is moving energy around, the mechanical behavior of materials that keep everything together and not falling apart or endangering people is very important. The types of breakage we're most familiar with are from tension, compression, shear, and torsion. A combination of tension and compression is produced when something is flexed. This is a very common loading situation, like a diving board when you're standing at the end. We see flexural stress in the pipe and duct between two hangers or dangling condensate lines. There are many other things that cause breakage, such as stress concentration, fatigue, impact, buckling, thermal expansion, and the always lingering problem of corrosion.

Strength

If you put a load on the material, it will stretch just like a rubber band. Materials have different values of elasticity; for example, steel is half as elastic as brass and 0.003 as elastic as most plastics. Nevertheless, all materials stretch or contract when stress is applied. If the stress is higher than what is called the *yield stress* of the material, most materials will start bending permanently. This is called *yielding*. Some brittle materials, like glass, don't yield; they just break right away. However, ductile materials like steel will stretch and thin out until they get to what is referred to as the *ultimate stress*. At the ultimate stress, a material breaks. With brittle materials the yield and ultimate stresses are almost the same. Fortunately, ductile materials give a warning before they break. They go from flexing and being elastic to permanently bending.

Shape

The size and the shape of a material are usually far more important than the yield stress. Big things can handle bigger loads. Cleverly designed shapes (like I-beams) are very efficient in handling big loads with much lighter weight.

The cross-sectional shape of a material radically affects how the material behaves under load. This is most obvious when considering I-beams. They are oriented so that they can hold the vertical load of a piece of equipment against gravity. I-beams are very strong and rigid in one axis and much less so in the other axes. If you know which direction a load is applied, you can orient a material advantageously. If the direction of loading is unpredictable, it's best to use round cross-sectional shapes such as tubes and rods because they handle loads equally in all directions. This is why your bones are basically tubes. You don't want an arm bone to suddenly break if you carry something in the wrong axis. Talk about adding complexity to your life! However, using I-beams and other specifically designed cross sections lets you use materials more carefully—which is good for a lot of other reasons.

Because you deal with structures, it is good to have insight into this behavior. When you make a penetration through a wall, you are thinking about the effect of smoke and fire migration. When you modify a material, you need to think about how it is weakening the structure. To destroy a beam's strength, drill a hole or cut a notch near the side under tension. I don't even like to use fasteners on the tension-loaded side. Adhesives are better. Moving supports to the middle of the beam where there is no stress is also a great method. This doesn't mean you can't use beam clamps, because beam clamps don't cut the beam web. To make a penetration through a structural support that minimizes the weakening, drill or cut through the middle of the beam.

Figure 5.1 shows an example of structural engineering requirements for beam penetrations. This diagram highlights the importance of leaving the highly stressed upper and lower portion of the beam untouched.

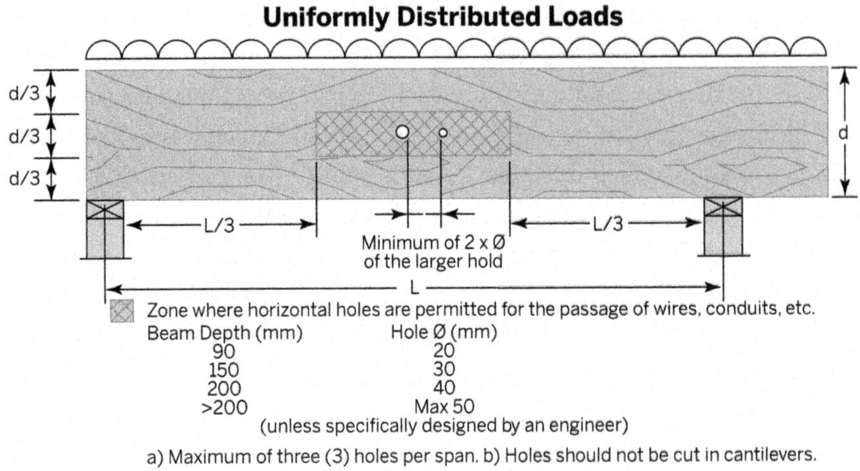

FIGURE 5.1 Recommendations for locating penetrations through a beam. (*Source:* SmartFrame by Tillin)

We will look at corrosion later, but keep in mind that corrosion causes things to break by reducing the area that can handle a load. Corrosion in areas of high tensile stress is particularly damaging.

SAFETY SUMMARY
Shape

Engineers carefully design structural pieces. Saws easily cut straight lines. One time I saw a TJI floor beam (or I-joist) where the bottom flange was cut out to make room for a P-trap. This quick action by a plumber with a saw destroyed the strength of the TJI by eliminating the part of the beam that carries the most tensile loading—the bottom of the TJI. The tensile stress in a beam is at its maximum the farther away the material is from the center (typically). The saw-slinging plumber cut exactly in the wrong place to fit his P-trap. Avoiding sharp edges and carefully locating where to drill holes will make a part much stronger than haphazard approaches. What do you do when there is interference? This happens all the time, sorry, but structure beats out plumbing.

Stress Concentration

A common factor in how many objects break is called *stress concentration.* A stress concentration is the amplification of stresses where an object's shape changes abruptly, such as a sharp corner or a drilled hole. You've probably noticed that cracks often originate at these types of features. The amount of stress amplification produced at these changes in shape is determined experimentally, but they can easily double the stresses in an area. That can be just enough to start cracks forming. Stress concentrations are reduced by blending shapes as best as possible, such as using large fillets. Ductile materials handle stress concentration better than brittle materials. Maybe now you are just a little bit worried about the sharp biting cut a beam clamp screw introduces into an I-beam web? Welcome to the engineering world of worry!

Another important note on shapes is remembering that long, slender objects change all the rules. These shapes can buckle at stresses much lower than the material yield stress. Using a long stick of pipe alone as a column for structural support is asking for trouble because of its high *slenderness ratio.* If you've been up on scaffolding, you know that buckling and torsional twisting are your biggest worries. Always think about these two things when dealing with structures.

Fatigue

When stress is applied and removed from a material, which is very common in the mechanical world, the material might break earlier than expected. This cyclic stressing produces a result called *fatigue.* Yes, this is the same word as what you feel like at the end of the day. However, in this case, it means the weakening of the material due to the growth of small internal cracks produced by stresses being applied and removed. Almost everything you deal with is vulnerable to fatigue failure because you're dealing with rotating fans and pumps and things that vibrate. The only things that aren't subjected to fatigue failure are objects that don't get these stress cycles. Fatigue is a big issue for things associated with rotating and vibrating equipment because the stress cycles count is much higher than something like your wrench handle that is only occasionally stress cycled.

You can see evidence of fatigue failure by looking at broken parts sometimes. You will see what are called *beach marks*, which are bands of different colored materials; or you may notice corrosion, where the material started cracking but was still strong enough to hold the load until the cracks grew further.

There is published data on the relationship between the number of stress cycles and strength. Typically, if the part is designed so the stresses don't exceed 50% of the yield stress, the material will last forever.

Corrosion

Steel pipes will often be coated to prevent rust. This coating, such as paint, keeps the oxygen from getting to the steel. Sometimes pipes are coated with zinc using a process called *galvanizing*. Not only does the zinc provide a physical barrier, but if the zinc is scratched, it acts like a sacrificial anode. A *sacrificial anode* is a material with loosely held electrons that are easily stripped away. This means the material will sacrificially corrode before the steel. This is the same concept as is used in hot water heaters where sacrificial anodes made from combinations of magnesium and zinc protect the tank.

Because corrosion is like a battery, where one material donates itself under the presence of another material, the material that more readily gives off electrons will corrode first. In the case of aluminum, copper, and stainless (or chromium) steels, when oxidation (like rusting in steel) occurs, the corrosion products bond tightly to the surface and provide a protective layer for the parent material. If the coating is worn or scratched away, a new coating quickly forms. In this way, these materials are protected from deep, penetrating corrosion. Some materials, such as low-chromium steel, have barely enough chromium to produce this protective oxide layer and can actually corrode through in caustic environments. For example, corrugated stainless steel tubing (CSST) used in low-pressure gas systems is made from low-chromium, AISI 304 stainless steel. Because it has a very thin wall, it is vulnerable to pitting corrosion.

Materials that have loosely bound electrons are normally good conductors of both heat and electricity. Therefore, metals are usually much more thermally conductive than plastics. Copper has better thermal conductivity than aluminum, which in turn is much better than steel. Silver is even better than copper, but it's soft and expensive. Gold fits in between copper and aluminum in terms of thermal conductivity. Imagine what would happen if heat exchangers were made from silver and gold!

Galvanic Series

Let's look a little more at the chemistry of corrosion. Corrosion is produced by electrons flowing, just like they do in a battery. The electrons in metals are loosely bound to the nucleus and are easily moved out of their orbits. If one metal has a property of wanting to take in loose electrons, it's called a *cathode*. If it wants to give them up easily, it's called an *anode*. The relationship between metals is shown in the *galvanic series*, as shown in Figure 5.2.

Titanium	*Most Cathodic (less corrosive)*
Stainless Steel (316)	
Stainless Steel (304)	
Nickel	
Copper	
Tin	
Cast Irons	
Carbon Steels	
Aluminum	
Zinc	
Magnesium	*Most Anodic (more corrosive)*

FIGURE 5.2 Galvanic series of common alloy groups.

Metals such as magnesium, zinc, and aluminum are anodic and want to give up electrons. Metals such as nickel and copper alloys are the opposite. They are cathodic and want to take in electrons. If you put these metals together, you have a collection of metals that want to make electrons flow. This creates current and corrosion. As the electrons flow, the material wastes away.

For example, zinc (Zn) wants to quickly shed two of its electrons, and as it does so, it becomes a positive ion. The way this looks in a chemical reaction is:

$$Zn \rightarrow Zn^{2+} + 2^{e}$$

This formula shows one zinc atom turning into one zinc ion with a positive charge and two free electrons with negative charges. Zinc is more eager to do this than any metal besides magnesium. The lost electrons are looking for a home, like iron, copper, or anything else on the cathodic side of the galvanic series.

The rusting of iron (Fe) in steel looks like this:

$$Fe^{3+} + 3(OH)^{-} \rightarrow Fe(OH)_3$$

This formula shows that one iron ion combines with three hydroxyl ions (produced from water) to become one rust molecule, $Fe(OH)_3$. As we discussed previously, many materials create oxide coatings that protect them from further corrosion. This process is called *passivation*, and the metals are *passivated*. Passivation occurs in common HVAC materials like copper, aluminum, and stainless (chromium) steels.

There are other important types of corrosion that can't be easily seen, such as *crevice corrosion*. Crevice corrosion occurs in hidden places, like fasteners, due to variation in electrolyte concentration. *Electrolytes* are liquids that allow current to flow. You can mix aluminum and copper in baseboards because there is no electrolyte. Stick the baseboard in a bucket of salt water and watch the aluminum plates get eaten away. I did this a long time ago, just to prove it to myself. You can see this in equipment located near the ocean because salt water is an excellent electrolyte. Fresh water is not very good, and air is really bad, which is good! Think about that weird statement for a while . . .

Other forms of corrosion include *pitting, intergranular* (which can also cause *stress corrosion cracking*), and *biological*. These are interesting but are probably uncommon concerns to deal with in your day-to-day life.

Figure 5.3 shows the result of a copper and aluminum connection. To be fair, this baseboard corrosion was accelerated by salt water being present that acted like an electrolyte.

FIGURE 5.3 Aluminum baseboard fins corroding in the presence of copper tubing.

CORROSION SUMMARIES

Steel

This one we have all seen. Without painting or galvanizing, the unprotected steel rapidly rusts. The rust is oxidized iron that flakes off and exposes more underlying steel. The oxide layer of cast iron tends to bond to the material, acting like a rust paint that protects the underlying iron.

Stainless Steels

Stainless steels are very resistant to corrosion, but their resistance depends on how much chromium and nickel are in the alloy. The chromium oxide layer protects the underlying steel from rusting. Nickel is great, because it never rusts. Some low-grade stainless steels, like AISI 303 and 304, are vulnerable to corrosion, especially when exposed to chemicals like muriatic acid and other solvents.

Copper and Brass

Copper and brass corrode in the presence of steel and aluminum in a moist environment. The corrosion can appear in different forms, from a rough green to a smooth brown. Brass can also be weakened by an effect called *dezincification*, where the zinc is leached from the material, leaving a bright copper color behind. Brass will also react with ammonia to produce a blue-green corrosion. Corrosion of copper electrical wire reduces the conducting area and therefore increases the resistance of the wire. Many of us have seen what that does. We are fortunate to have circuit breakers.

Aluminum

Aluminum is very vulnerable to galvanic corrosion because it is at the wrong end of the galvanic series compared with other materials. It will readily react with copper in a moist environment. In a dry environment, such as baseboard heat exchangers where aluminum plates are physically attached to copper piping, this isn't a common problem. The product of corrosion is a crusty, light gray scale.

Thermal Expansion

Materials expand when they are heated and shrink when they are cooled. The amount of shrinkage is a function of the material, the temperature change, and the length of the pipe or duct. If the material is not allowed to expand or contract, internal stresses will develop.

In piping systems, thermal expansion can be handled by making the pipe system flexible so it can move around as it expands and contracts. However, thermal expansion is also accommodated with bellows and coaxial expansion joints.

You have to be especially careful that you consider different materials and how they will grow or shrink with respect to each other. This can be found by looking at their *thermal expansion coefficient*. For example, copper expands roughly one-and-half times more than steel. Aluminum has twice the thermal expansion rate of steel. Plastics grow a lot more than metals. HDPE expands about six times more than steel. Plastic pipe expansion rates vary a lot depending on their specific composition. Fortunately, some plastic pipes are viscoelastic, meaning they sort of flow like fluids and can therefore handle a little thermal expansion internally.

You need to get information on the thermal expansion coefficient for your specific materials—they can vary a lot. For example, the term *polyethylene* does not cover the wide range of values associated with these plastics.

The change in length is based on the change in temperature multiplied by the material's coefficient of thermal expansion. This is then multiplied by the initial length of the pipe or duct—short things don't expand as much as long things. For example, if 100 feet of steel pipe is heated up by 130 °F, it will expand 1 inch.

Thermal expansion values vary based on specific material, but be warned, plastic grows a lot when heated. The thermal expansion coefficient indicates the amount of thermal expansion for every degree of heating. These values show that polyethylene grows almost 13 times more than carbon steel. Try keeping your hot water PEX tubing straight with those numbers!

While plastics are generally viscoelastic, it is vital to follow manufacturers' recommendations on how to handle thermal expansion. This is a very well understood phenomenon, and there are many practical approaches to handling it—viscoelasticity isn't one of them, although it is a fun word to say during a lull in conversation.

SUMMARY OF WORRIES

Here are some things you should always be worrying about regarding materials:

1. Noncorrosive materials can corrode (with some exceptions like nickel and gold—and good luck with that specification requirement!).
2. Materials can interact with each other to corrode, such as copper and aluminum.
3. Crevice corrosion occurs in areas you can't see, like threads.
4. A part can be working fine and then suddenly fail due to fatigue.
5. Structural pieces can be oriented in the wrong way, which makes them weak.
6. People cut and drill things haphazardly, thereby weakening them.
7. Plastic will create smoke when heated.
8. Plastic will get weak at high and low temperatures. Metals get weak at extreme temperatures also, but their range is much higher.

SIX

CORNY STORIES

Corny Story 1

Combustion and Fluid Flow

Let me illustrate what we have been discussing about combustion and moving heated water from the boiler to the air handler with a little story.

Imagine you're a hydrogen molecule. Imagine further that an oxygen molecule snuggles up next to you. Suddenly you hear a clicking and see sparks. Waves of heat start flying out of you as you join up and link with the oxygen molecule—a shotgun wedding. Now you've become a molecule of water vapor.

You're flying through the boiler, and you look over at some molecules flying next to you and say, "At least I'm doing something, unlike those nitrogen dudes. They're just going along for the ride. Stealing my heat and doing nothing for it."

You fly past some tubes, which are cold. You give them a lot of your heat. Now some engineer is going to give you just enough velocity to spit you out the exhaust flue, make a little vapor cloud of glory, and end up in the so-called water cycle, which means you'll be flushing someone's toilet soon.

Now let's say you magically come back. But now you are a hot, liquid water molecule moving away from the boiler in a nice cozy pipe. It's you and the other water molecules, a lot simpler than the boiler mob. You're at peace and say to yourself, "I'm moving with my buddies at a nice easy pace—feeling nice and toasty." But your mellow thoughts are broken as you look down at the pipe wall with disgust. "What are those guys doing down there?" you ask, and then say, "They're just loafing!" No one responds, and you think for a minute, "Oh yeah, I guess we're going kind of slow out here in the middle of the pipe too. I guess our Reynolds number is way below 2,000. We're just going laminar with a thick, slow boundary layer and no drama."

Suddenly things start changing—your velocity starts going up, and you're getting squeezed into something tight as you approach the air handler and its heating coil. You note, "I think my Reynolds number is getting close to 2,000. I can feel it in my watery bones." You hit some ridges and start tumbling around and get squeezed up next to a metal wall. "Brr, that metal is cool," you say, missing your younger days when you were hot and cozy. "Oh boy," you exclaim, "this low Reynolds heat exchanger has little ridges, and they are tripping me into a turbulent boundary layer! Now we're all squeezed up into basically nothing, and that metal is sucking all the heat out of me!"

After a painfully long time of dragging across the metal, you feel cooler, maybe only 12° or 15° cooler, but it's noticeable. At the end of the air handler ride, you slow down again and things are like they were. You're with the same buddies, but you're all going in the other direction and everyone else is a lot cooler too.

"We're headed back to the boiler," one of the old-timers yells with the assurance of someone who has done this many times before.

Corny Stories 2 and 3: The Refrigeration Cycle—a Different Way to Explain It . . .

I'm going to go over the refrigeration cycle with two stories. We talked about refrigeration a while back as controlling temperature with phase, but let's dig in a little bit with these silly stories. If you don't like them, go find a thick book with lots of diagrams or an animated video. I like my corny stories . . .

Note in these refrigeration stories that (1) bounciness represents temperature and (2) hand magnets represent molecular surface forces.

Corny Story 2

Boiling Water

You and your friend Pat find yourselves in a pan sitting on top of a stove. Like most friends, you have magnets attached to your hands that stick your hands and Pat's hands together if you get close to each other. As it turns out, you and Pat are really cold. So cold that you don't bounce around at all. You just sit there with the atmosphere pushing on your heads. This is enough pressure to make both your and Pat's hand magnets stick together. Egads, you are solid ice!

Suddenly some wise guy decides to turn the stove on, and you and Pat start bouncing around from the heat. You and Pat get warmer and warmer while still stuck together with both hand magnets. Suddenly, you find yourself vibrating so much that you and Pat are breaking free at one hand. You are partially free! You are now a liquid. You and Pat celebrate with the one-hand-linked-wiggle-liquid dance.

Wait! You are now getting hotter and hotter. You bounce around more and more. The one magnet holding the two of you together is having a hard time keeping you stuck together as you do your wild dance. Finally, it's too much, and the magnets can't hold you and Pat together anymore. Pat is bouncing around and moving away from you, and all the while you are in a frenzied mode too. This vibration is too much for your hand magnets to overcome. You are now a gas. Congratulations, I guess.

Then the same guy that turned on the heat decides to put a tightly fitting lid on your pot so nothing can escape from it. You look down and see that some of your friends are still liquid water, but you are above them in the gas phase. Now you are starting to get squeezed tightly together with other guys, even though you are still bouncing around like mad. As you get squeezed closer and closer, the magnets become close enough that they start sticking again. You see this happen to your neighbors, and they turn into a liquid. If only you could find Pat in this crowd! You are getting so crowded that you stick to a molecule and turn into a liquid too. As you turn to look, you see that it is Pat and feel some peace because of it.

Then the guy decides to take the lid off again. Suddenly, there is no pressure. You are bouncing around just like you were before, but freed from the crush of that sealing pot lid—whoosh, you bounce away from Pat, and you are both gases again.

As you fly into the guy's face, you are amazed at how either pressure or temperature can make you change phase. You smack the guy's face as high-temperature steam. You cool down and link up with Pat again as a liquid and then flow down to the very end of that guy's pointy chin.

Corny Story 3

Walking Through the Refrigeration Cycle

Imagine you and your friend Pat are two refrigerant molecules going through a refrigerator. As in Story 2, you and Pat are equipped with the latest hand magnets that stick the two of you together if you get close.

Let's say you are about to enter the compressor. At this point you and Pat are vibrating so much due to your giggling (oops, I meant temperature) that you bounce off each other—neither of your hand magnets are strong enough to overcome your bouncing around. You're a gas.

The compressor's orbiting scroll starts opening, and suddenly you and your friend are sucked into the compressor. As you wait to see what happens next, the scroll starts squeezing you closer to Pat so that the magnets are trying to get you to stick together, but you have a little too much bounciness and are still bouncing off each other. You are pushed out the discharge of the compressor, and suddenly you are traveling through a coil. You are hotter and bouncier than the copper coil and the air outside it. You start to give up your bounciness to the coil and air outside the coil. You are cooling down, and the outside air is heating up. This cooling down reduces your bounciness, and you start sticking to Pat due to those hand magnets. That means you have become a liquid. You gave all that potential energy of the magnets to the air and made it hotter.

Now you are a cool, high-pressure liquid; but then you and Pat are pushed toward this tiny, tiny tube. You squeeze through at a really high speed, losing all sorts of energy in the form of pressure until you get out on the other side into another coil under low pressure. Low pressure means you and Pat can bounce around easier—you're not being squeezed together. You have so much freedom now that your magnets aren't being pushed close together. Therefore, you are not sticking much and you can become a gas. That nasty pressure had been forcing you together against your will so you had to be a liquid, but now you are free to be a gas—you just need a little energy from the air and you can break free of the magnet's stickiness. You take the heat from the outside air, which cools the air, and you happily go back to the way you were before that compressor sucked you in.

But wait, as you and Pat are doing the "I'm a gas" victory dance, you get sucked into the compressor again. You yell out, "Oh no! Here we go again!" Another refrigeration cycle.

Now for all you hydronic and ductless mini-split folks, you can leave. But for the rest of you who push air around a building, hold tight for Section 2. We'll pick it up from the air side of the air handler until supply air is placed precisely in the conditioned space. The good news is most of the hard-core engineering concepts have already been introduced. It's downhill from here.

SECTION 2

THE AIR SIDE

FROM THE AIR HANDLING UNIT TO THE CONDITIONED SPACE

SEVEN

CONDITIONING AIR

We will start our deeper investigation of conditioning air by considering operations in the context of the normal goal of HVAC systems, namely human comfort and safety. After this presentation, we will look at the function of air handlers.

Comfort and Safety

One of the main goals of HVAC systems is to provide comfort and safety. Safety involves reducing fire hazards by pressurizing exit routes, reducing paths for smoke migration, providing fire stops, and using materials that do not burn or create much smoke when hot. Modern HVAC systems monitor carbon monoxide and other hazards. For comfort conditioning, we primarily consider temperature and humidity. However, there is more to it. For example, radiation can offset coolness, and draftiness can offset heat. Therefore, we don't mind if it is warm and windy, nor do we mind if it is cool and sunny. We are sort of boxed in by these four values: high and low temperature, high and low humidity, high and low radiation, and high and low draftiness. These are all affected by what we are wearing and how active we are. For most indoor conditions, we are pretty comfortable without any wind and radiation as long as the temperature and humidity are satisfactory. ASHRAE Standard 55 (Thermal Environmental Conditions for Human Occupancy) provides commonly acceptable temperature and humidity values for different seasons.

In addition to temperature, humidity, radiation, and draftiness, we consider other factors, such as noise, vibration, carbon dioxide, and all other indoor air quality issues. Carbon dioxide (or stuffiness) is one of the most notorious problems. Building design has gone from being naturally drafty and inefficient to overly sealed up and efficient. Now we are trying to control exactly how much outside air is allowed to enter a building. There is a comfort zone that you try to produce in which the temperature and humidity relationships are acceptable to most people. In addition, you try to keep the air fresh by bringing in outside air and monitoring carbon dioxide.

For constant volume systems, you size the ducts to keep velocities constant to the air diffusers. This is a peculiar mixture of making ducts smaller and cheaper versus standardizing duct sizes for construction efficiency. The math on this is pretty easy. The velocity is equal to the air volume divided by the area of the duct. Velocity is a very important number in HVAC design, which we discussed previously regarding piping systems. Design guidance is available for the optimal velocity across heating and cooling coils, duct fittings, and other ductwork. Your thermal calculations tell you how much air volume you have to deliver, and these velocity guides tell you what is acceptable.

Air distribution is intended to deliver supply air and distribute it uniformly through occupied zones. However, you want the air to be delivered quietly and mix well so that the whole space is comfortable. You don't want to create high-speed airflow that annoys people and makes them hate you because they are getting cold supply air blowing on them, making them freeze on a hot summer day while messing up their hair and sounding like an approaching airplane. ASHRAE Standard 55 provides details. It generally recommends velocities under 50 feet per minute (fpm) in the facial region as acceptable to most people.

People generally accept an air distribution's white noise, which is diffuse noise over a wide range of frequencies. What they don't like are loud noises, especially in the high-frequency spectrum where it is especially annoying. No one likes screaming fan belts, but they also don't like the sound of your fans spooling up, ducts popping, and giant whooshing sounds. Okay, those are still better than the fan belt screaming.

Decibel scales weigh the high-frequency component of noise heavier than the low-frequency sounds. So a sound that has the screaming high-frequency noise will have a higher decibel rating than the same sound power delivered at a lower frequency. While HVAC systems normally don't exceed OSHA standards on noise, they do produce the background noise of many buildings. ASHRAE uses noise criteria (NC) or room criteria (RC) values. The recommended values range for office spaces from a low of NC/RC 25 for private offices to 45 for open offices. The values for computer and mechanical areas can be much higher. Table 7.1 provides representative noise levels for various activities and sources.

TABLE 7.1 Common Sound Decibel Levels

Sound Source	Decibel Level (dB)	Approximate Perception	Potential Effect
Quiet room/breathing	Below 20	Faint	Threshold of hearing
Whisper/rustling leaves	30	Very quiet	Non-disturbing
Refrigerator hum, library	40	Quiet	Manageable
Electric fan	50	Moderate	Pleasant background
Normal conversation/talking	60	Moderate	Typical speech level
Vacuum cleaner/dishwasher	70	Loud	Noisy
Traffic	80	Loud	Noisy
Lawn mover/motorcycle	90	Very loud	Eardrum damage after 8 hours exposure
Rock concert/chainsaw	100	Very loud	Eardrum damage after 15 mins exposure
Jet engine taking off	Above 140	Extremely loud	Eardrum damage without protection

COMFORT AND SAFETY CONCERNS

Comfort associated with HVAC can be considered a combination of:

- Temperature
- Humidity
- Air velocity
- Radiant surface temperatures
- Noise
- Vibration
- Carbon dioxide level
- Air quality

All these factors are important and are carefully covered by a lot of standards. ASHRAE and ACCA (Air Conditioning Contractors of America) do a good job of pinning these down. These types of lists grow longer and more complex, especially when dealing with something as vague as comfort. Someone will always want to add something. Feel free to do so.

Air Handler Units (AHUs)

We left the HVAC process at the water side of the air handler. Now let's move over to the air side. We will now look at the part of the heating and cooling diagram where the heated or chilled water is brought to an air handler, as shown in Figure 7.1. This figure is extracted from the whole schematic of the boiler and chiller previously provided in Figures 1.3 and 1.6. Figure 7.2 shows a typical configuration of an air handler.

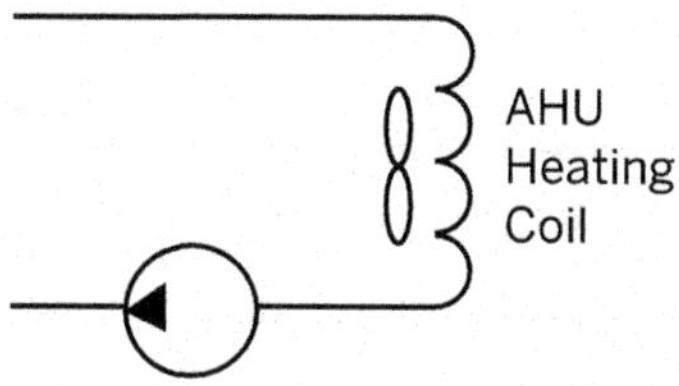

FIGURE 7.1 Air handler section of the heating and cooling schematic.

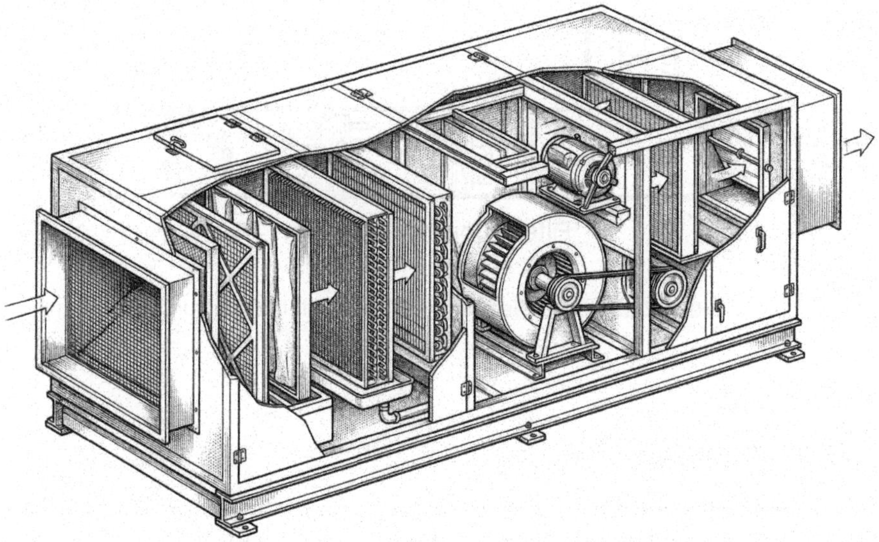

FIGURE 7.2 Layout of a typical air handling unit.

Respectively, hot or chilled water is piped from a boiler or chiller into the air handler to provide heating or cooling. The water goes into a heat exchanger, and air is blown through the heat exchanger so that it picks up either the heat or coldness of the water. From the air handler, the conditioned air typically gets ducted to a diffuser and pushed into a space. Notice, I use the word *typically* a lot—I recognize there are a lot of ways to move energy around.

Figure 7.3 shows a simplified schematic of an air handler that reveals the main parts and pieces that let it control such things as air temperature, humidity, quality, and space pressurization. Air handlers must respond rapidly to room conditions. The control systems that operate the dampers, coil flow rate, and fan pressures are always dancing with all the sensing devices that keep people safe and comfortable.

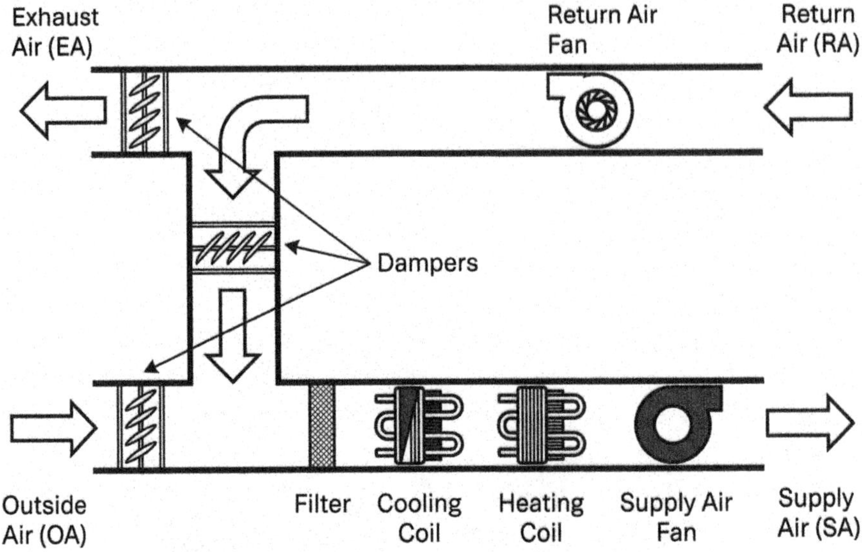

FIGURE 7.3 Simplified schematic of an air handling unit.

Conditioning Spaces

Air handlers are in charge of getting the supply air to the desired location. They send supply air to *conditioned spaces*, which is the general name for an enclosed room. Air handlers also control humidity, filtration, carbon dioxide level, duct pressure, draftiness, and noise. In addition to heating and cooling the supply air, the air handler takes air in from outside to ensure that fresh, low carbon dioxide air goes to the people in the conditioned space. Finally, the air handler manages the return and exhaust air to maintain a desired pressurization in the space. The air handler does this using dedicated fans and/or dampers.

Pressurization

Spaces can be positive, neutral, or negatively pressurized, depending on what the space is used for. Positive pressure will ensure that air only flows out of the space.

Positive pressure is desired if you are concerned with all the air coming in being tightly controlled, just as it would be in a surgical suite or clean room. Negative pressure will ensure air only flows into the space, just as if you were dealing with hazardous chemicals or airborne infections and you didn't want the bad stuff to escape unfiltered. The engineering concept is pretty simple. Push more air into a space than you remove and it will be positively pressurized. Suck more air out than you put in and it will be negatively pressurized.

There are different classification systems for the pressurization standards, such as ISO classes. In the healthcare industry there are rooms designated *Class Q*, which, in addition to negative pressure, have features to ensure that infectious diseases don't escape the room. These quarantine rooms include things like alarms, self-closing doors, and anterooms.

Coils and Humidity Control

Because we're dealing with temperature change, we deal with condensation all the time. As mentioned previously in the Corrosion section in Chapter 5, many metals will corrode due to moisture. Because we sometimes create condensation, HVAC technicians should know where it might form and settle. You also must deal with how outdoor weather affects operating systems. Psychrometric charts let you plot how temperature and moisture content vary as air is mixed. However, generally, we are very concerned with outdoor air moisture because it decreases the performance of certain functions like cooling towers and swamp coolers.

In order to get the right heating and cooling of supply air, the heating and cooling coils have to be designed so they are both thick enough and dense enough to be effective. In addition, the air and water have to pass through at the right speed. Cooling requires dehumidification in the coil, so the cooling coil has to be thick enough, and the air velocity low enough, so that the liquid water condensate can drop out in the air handler. Figure 7.4 shows humid air passing through a coil and the relationship between velocity and condensation.

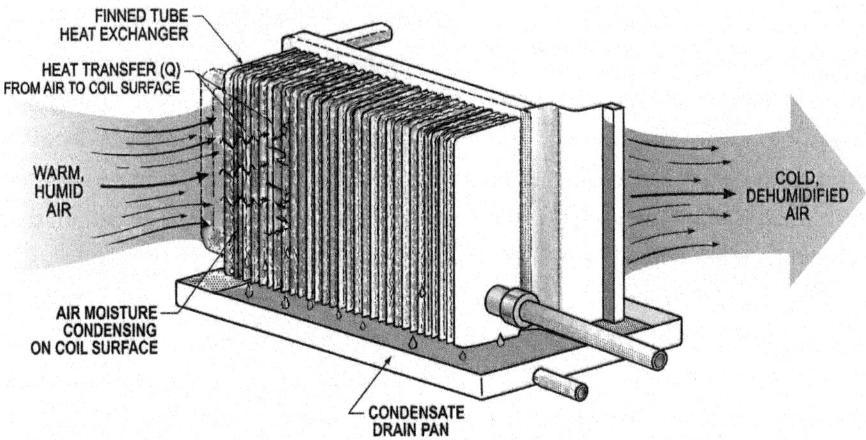

FIGURE 7.4 Water condenses as it passes through coils. The velocity needs to be low enough that the water doesn't carry over into the cooled discharge air.

Sometimes desiccant wheels are used to dehumidify. These work by having the air run through a material that is capable of absorbing water vapor. Note that desiccant wheels are similar to total energy (or enthalpy) wheels. However, total energy wheels work between the outside air and supply air, like a mass transport system. Psychrometric charts are a great way to visualize these processes.

Fan Inlet and Discharge

The design of the ductwork at the inlet and discharge of a fan is very important. The details are well understood, but if these inlet and discharge geometries are not ideal, the fan will always be inefficient, no matter what else you do. The flow behavior right after a fan discharge is very jumbled. It takes a while for the sticky, viscous forces of interacting air to slow down the fast molecules and speed up the slow ones. Therefore, it is important to have a smooth, straight transition at this section; otherwise some of the air will be moving at a very high velocity and some air might be stagnant. Any branching off from the duct has to be done after the air is fully developed. *Fully developed* means the air velocity is constant, with a higher-speed core and a slower-speed boundary layer touching the duct walls. Figure 7.5 illustrates this development. The fan inlet is also important, so follow the rules on all this stuff. Let the fan breathe! Pumps have similar design guidelines, but fans need a more carefully designed inlet and discharge lengths. I'll put in the calculation for this transitional duct length in the Appendix because poor transitions in fan discharge are the biggest design oversights I see.

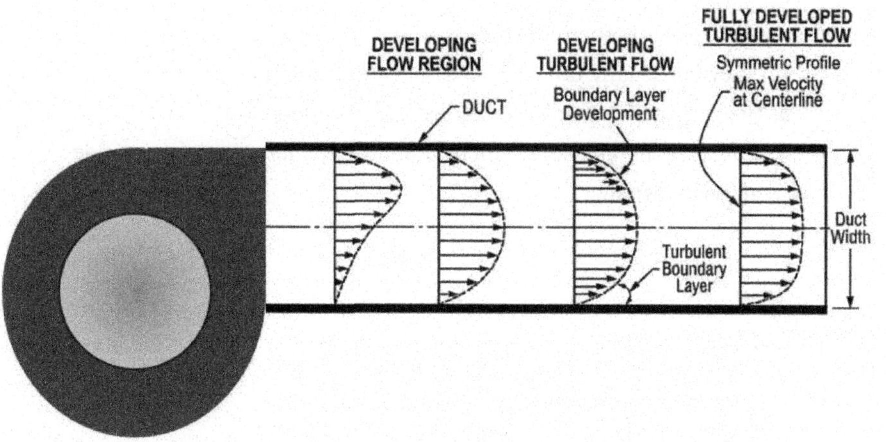

FIGURE 7.5 Development of flow in a duct after the fan discharge.

The airflow through coils is also important. You waste a lot of coil effectiveness if you have different flow velocities in a coil. If high-velocity flow is going through a small part of the coil, not only does it *not* exchange heat as well, but it can also allow for condensing moisture to carry over into the duct airflow. The inlet to coils should be gradual so that air flows uniformly through the coil. A perforated diffuser plate can also be used to provide uniform velocity, but this introduces a pressure drop that consumes fan energy.

SUMMARY OF WORRIES

Here are some things you should always be worrying about regarding air handlers:

1. They are the soul of HVAC systems, but they are big and ugly, and no one wants them taking up space.
2. They are usually not designed for ease of maintenance.
3. Sealing them completely and reliably is like herding cats.
4. Everyone hates making membrane roof penetrations for outside and exhaust air.

EIGHT

MOVING AIR

Ducts

Ductwork is different than piping in one important way—it's an open system. This means that you start off with high-pressure air at the air handler, but the pressure keeps dropping until it gets to almost atmospheric pressure at the air diffuser/terminal device. Yes, it gets drawn back through the return ductwork, but there is a disconnect in the air system that you don't get in the hot and chilled water piping systems. You may recognize that the piping to and from the cooling tower is also an open system.

The basic goal of duct sizing is to ensure that the velocity is constant, even when supply air is being tapped off. This is done by reducing the duct size to keep the duct velocity constant. The relationship between velocity, flow, and duct area is the same as what we discussed previously, namely area equals flow rate divided by velocity. Algebraically, this means velocity equals flow divided by area. Therefore, as the flow is reduced, because you're branching supply air, the duct area has to decrease to keep the same velocity.

Constant Air Volume and Variable Air Volume Designs

Balancing the pressure drops to all the diffusers is important in constant air volume (CAV) systems. This balancing can be done with dampers, but an elegant design reduces the need to choke off some ducts with an inefficient damper by making the duct smaller. A smaller duct will build in the pressure drop, all the while being physically smaller and cheaper. It is very unusual to design "inefficiency" into a system, but this is one example. Figure 8.1 illustrates a CAV with four ceiling diffusers and one return.

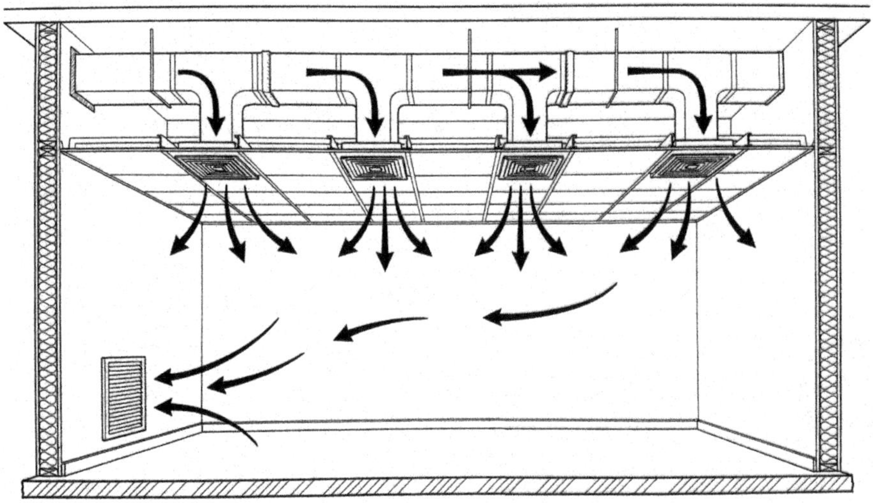

FIGURE 8.1 Constant air volume system.

Variable air volume (VAV) systems don't require the same care in balancing the system. As long as there is enough pressure at the VAV, it will control things from there. Fitting designs that are gradual and use turning vanes provide a lower pressure drop and therefore more efficiency. Most duct systems are considered low-velocity systems. High-velocity systems are used where the duct size needs to be small. High-velocity systems have higher pressure drops and more noise.

The engineering goal of the duct system is to deliver the right pressure to the diffusers or VAVs so they can perform correctly. For constant volume systems, the system must be carefully balanced so the pressure drops are equal and none of the air distribution systems steal supply air from other sections. If you drill a big hole at the spigot end of your garden hose, all the water will go out the hole near the spigot and very little will go to the end of the hose.

If the system is not balanced, you have to start closing your balancing dampers. Partially closed dampers waste money. If you design a balanced system by appropriately downsizing the duct, you don't need to introduce a big pressure drop with a balancing damper. Smaller ducts are cheaper and easier to install, which saves money. Of course, there is a practical benefit in standardizing duct sizes, so you don't want to get too zealous if you are working with short runs.

Static and Velocity Pressure

There are two types of pressure that we consider in ductwork, and they both derive from the first law of thermodynamics, described previously. In duct design, the terms are unique. The pressure pushing out on the duct is called *static pressure*, and the velocity in the core of the duct is called *velocity pressure*. Velocity pressure is the pressure produced if you suddenly shut off the duct, just like what happens if you shut off a fire and smoke damper. This is the same idea as surge pressure in liquid systems. These two pressures combine to give the total pressure. You are playing between potential energy (static pressure) and kinetic energy (velocity pressure), as described in the first law of thermodynamics (from which the Bernoulli equation is derived)—velocity is equal to pressure, and pressure is equal to velocity. You can change the type of energy by duct design. At the diffuser, you are turning the total pressure into air movement, which is kinetic energy.

Duct Efficiency

Unlike water and fuel, the air in ducts can compress and cause pressures to change a lot. Because ducts can get very big, turning vanes and splitters are used to reduce the bunching up of air around corners due to the inertia of the air wanting to keep the air molecules moving straight. Inertia is a property of matter—matter doesn't like to change direction or speed. Newton's law of motion states that objects at rest want to stay at rest, while objects in motion want to stay in motion and in a straight line and at a constant speed. You have to use a force to make them change.

You can look up a duct fitting's pressure drop and see the exact loss associated with different designs and velocities. It is sad to see poorly designed duct systems. These systems require a lot of fan energy to get them to distribute air well—and this energy will be required for the life of the system. It is wasted energy. Sometimes the building's constraints require tight elbows, squeezes, and short transition sections, but these are sad to witness. When space is at a premium, the use of water and refrigerant pipes to transmit thermal energy is especially appealing because they require much less space than ductwork.

Designed Inefficiency

HVAC piping and ductwork are meant to be in service for decades. So bad designs waste a lot of energy over their lifetime. One of the saddest designs I have seen was a long run of commercial ductwork that started with a narrow squeeze over a girder near the discharge of a huge air handler. After the squeeze, an inline axial fan was added, probably a retrofit to make up for the pressure loss of the squeeze. For the life of this duct system, the squeeze was going to produce tremendous pressure drops, and that silly axial fan Band-Aid solution was going to have to run all the time. This points to the problem of when the ductwork is the last consideration after the structural steel and piping. This is so sad, because while you can overcome the duct squeeze with fan power, nobody really seems to care about this energy inefficiency except those of us that look underneath buildings and around mechanical equipment rooms. Inefficiency by design is a nasty problem.

Straightforward designs can also be challenged by air distribution design. Figure 8.2 shows a long, straight section of duct with one return at the near end. The return is basically sucking all the air out of the first duct discharge collar so that the conditioned air doesn't travel to the end of the duct.

FIGURE 8.2 Supply air is short-circuited to a single return in the lower wall.

Fire and Smoke Dampers

Design and code requirements call for fire and smoke dampers in ducts. Ducts make huge wall penetrations and can be an easy path for smoke to travel through a building. Fast-acting, reliable fire and smoke dampers are vital in ductwork because they prevent this from happening. These dampers have a specific hourly fire resistance value. This design requirement for ductwork is so important that it gets its own subheading!

NFPA Standard 80 (Standard for Fire Doors and Other Opening Protectives) and NFPA Standard 90A (Standard for the Installation of Air Conditioning and Ventilation Systems) govern the performance of these dampers and the materials used in HVAC systems. The IMC (International Mechanical Code) guides where these systems should be installed, along with their accessibility, inspection, and maintenance. UL Standard 555 (Standard for Safety for Fire Dampers) and UL Standard 555S (Leakage Rated Dampers for Use in Smoke Control Systems) set requirements for certification of damper performance. Just as I'm grateful for the experts on pressure relief valves, I'm glad there are people dedicated to getting this right. Figure 8.3 shows a fire and smoke damper that has fusible links and smoke detectors and is integrated into the building management system (BMS).

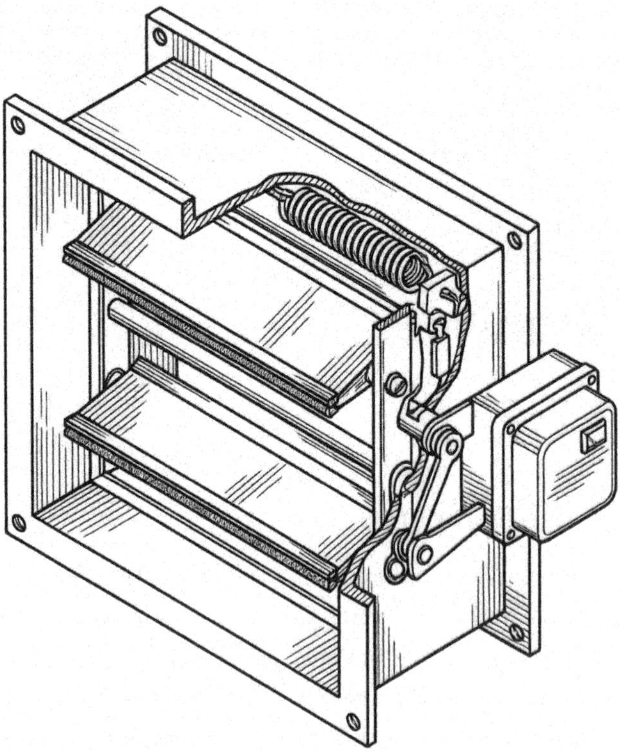

FIGURE 8.3 Smoke and fire damper.

DESIGN APPROACHES FOR DUCTS

Galvanized sheet metal is the most common material for commercial applications. Steel has a good strength-to-weight-to-cost ratio and will not produce smoke. The ducts can be rectangular, round, or oval. There is a wide array of standard fittings and custom fittings that can be developed by clever sheet metal gurus.

Ducts take up a huge amount of space, and they always seem to be last in priority for routing. Clever work is often done to make the best of difficult situations, whether it be snaking large ducts around other things, as shown in Figure 8.4, or running ducts outside of historic buildings and just showing off the glory of glinting sheet metal, as shown in Figure 8.5.

FIGURE 8.4 Ducts have to compete for space with many other mechanical systems.

FIGURE 8.5 Ducts are sometimes forced outside, such as with this historic stone building.

Fabric ducts are excellent in certain applications, especially large open areas. They are light and can have different types of air diffusion capability. They let you both duct air and diffuse it in one light, low-cost package. They have some drawbacks such as often being translucent and showing debris at the bottom. Because they are flexible, they can be a bit dynamic, flapping and sagging as they turn on and off.

Flexible ducts are made with a plastic liner surrounded by insulation. The liner is kept in position with a wire coil. These allow long, low-cost duct runs. They are physically weak and vulnerable to damage or leakage.

Fiberboard is in a similar category, and people have different opinions on it. Fiberboard is easily customizable and lets you make a duct system from material that fits in a sedan.

SUMMARY OF WORRIES

Here are some things you should always be worrying about regarding ducts:

1. They penetrate fire walls and can let smoke and fire spread.
2. They have the responsibility of delivering safe, conditioned air to occupants.
3. Sizing is usually dictated by the incomplete science of balancing velocity, cost, and standardization.
4. Giant swaths of sheet metal or duct board are hard to seal and insulate perfectly. Ducts are vulnerable to leakage and thermal conductivity losses and ongoing energy losses.
5. They can take up a huge amount of space, and even with BIM they have to make design efficiency compromises.
6. People can sneak around in them. Embassies and other secure locations have used security bars in their ductwork long before Hollywood shared that vulnerability with everyone.

NINE

AIR DISTRIBUTION

Air distribution has a flavor of both art and science. It relies on human factors, such as people's sensitivity to draftiness and noise. The values associated with these factors are not hard and fast numbers, but industry recommendations. The goal of the air distribution system is not only to mix the supply air with the room air, but also to ensure that the air does not feel drafty or noisy to the occupants. It's like pitching a ball and getting it exactly where you want over the plate. A designer wants to get the molecules at the wispy end of the supply airflow to hit the occupants at a certain speed—not too slow and not too fast.

Diffusers

There are lots of different terminal devices and ways of moving heat and cold around. I'm not trying to chase all of these down, but to consider general background information on diffusers. If you have read this far, you are curious, and I hope this discussion helps satisfy your curiosity. Figure 9.1 shows a typical ceiling diffuser. These diffusers terminate the ugly ductwork and present it as a pretty metal thing in the ceiling. They also reduce the speed of the airflow and spread it in a predictable manner. You can also hang party decorations from them.

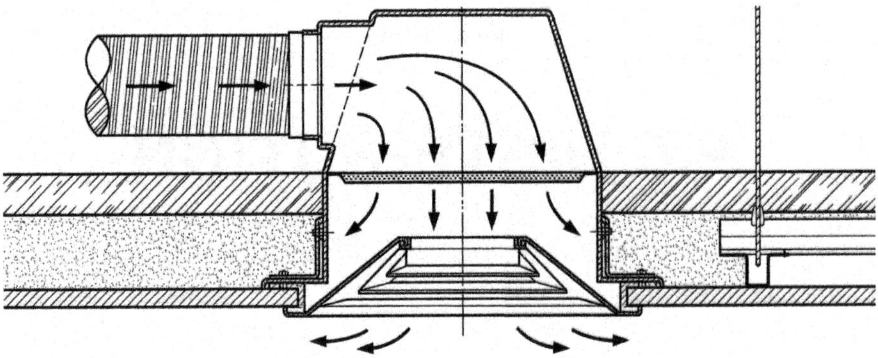

FIGURE 9.1 Typical ceiling diffuser.

Air diffuser manufacturers will give you numbers on a diffuser's radius of diffusion, throw distance, noise, and other types of measures that let you understand how this complex three-dimensional problem of blowing very cold (or hot) air into a space behaves. The human neck is one of the more sensitive areas for perception of air movement, and there is data associated with that metric.

What you are trying to do in most HVAC design is take all the noisy, clanky stuff in the mechanical equipment room and turn it into the whisper of air that is contacting a human and making that human feel comfortable. It's like throwing a perfect pitch from 1,000 feet away. The way the air hits human occupants is the strike zone. You want that air that you "designed" to hit people at the right temperature, humidity, carbon dioxide level, noise level, and velocity. Figure 9.2 provides an isometric view of the radius of diffusion for a ceiling diffuser. This is often given as a table of velocity versus distance.

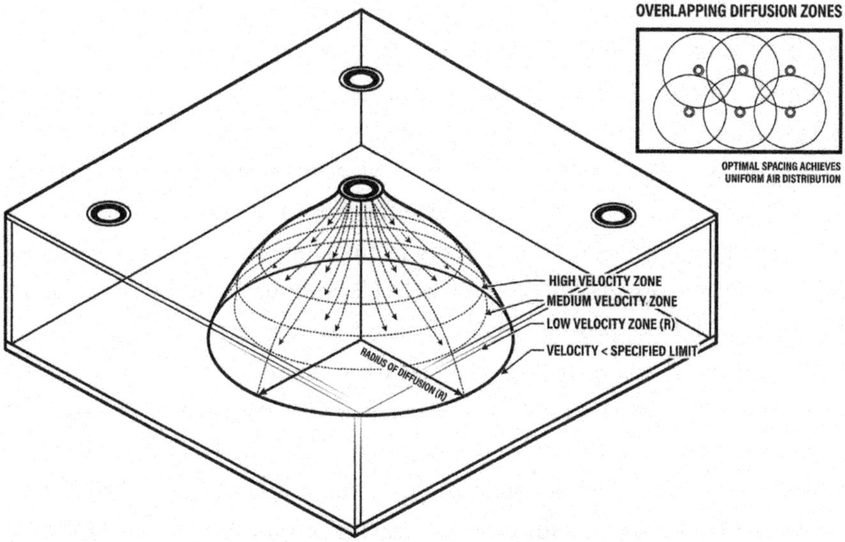

FIGURE 9.2 Radius of diffusion.

Behavior of Supply Air

Let's begin by looking at how air behaves in a room. First, we need to know that hot air doesn't rise. Yes, I'm being kind of geeky with that statement, but it's true. Hot air doesn't rise. Why would something rise under the pull of earth's gravity? Doesn't everything fall? What's really going on is dense air gets pulled down and pushes the low-density air upward—even under the force of gravity. Cold air is denser than hot air, so cold air sinks and forces the hot air to go upward against gravity. I guess you could call that "hot air rising," but its actually "dense gases settle." In practice, hot air is more buoyant with respect to the surrounding air. You can get into other issues here if you're dealing with other gases. For example, a helium balloon rises in air even if the helium is colder than the air. If you put a helium balloon on your passenger seat, you will see it move forward when you brake. This is caused by the denser atmospheric air moving backward toward the backseat and therefore pushing the less dense helium forward. It has nothing to do with temperature. It has to do with density and inertia. Buy a helium balloon sometime and try this out; it will make all these words stick.

We can take advantage of air density and the effect of moving air, and, yes, it's okay to think of hot air rising and cold air settling (but it really is density that's making this all happen under the force of gravity, not the temperature—I just wanted to say that one more time for fun). High temperature (or thermal motion) lowers the density of a gas. Molecules work between a balance of thermal motion and surface forces. If the surface forces are higher than the thermal motion, the molecules will stick together and become a liquid. If the thermal motion is high enough to overcome the molecular surface forces, the molecules will stay in the gas phase because the stickiness of the surface forces is insufficient to capture the fast molecules. It's like trying to catch a dribbling basketball with a piece of tape; the tape may momentarily stick to the ball, but the ball has too much inertia to stay stuck to the tape. But if the basketball is just barely bouncing, the tape will be strong enough to stick to it—the ball's inertia is less than the tape's stickiness. You may recall I used a hand magnet allegory in two of my corny stories. The sticky tape was just another way I thought of to illustrate surface forces. Surface forces explain a lot of the behavior of liquids and gases! You really want to understand them.

The behavior of the air in a room is a function of velocity, which is produced by the pressure in the ducts and the temperature differences. Diffusers (or other terminal devices) can be designed to push air in different directions. You can corral air to go wherever you want by working with velocity and direction of flow and understanding buoyancy and the effect of density.

Air Viscosity and Mixing

The viscosity of air (which you can't change) is produced by the surface forces of the molecules—their desire to stick to things. Air wants to grab other air molecules and drag them along in the supply airflow. This is the same thing we discussed when we looked at boundary layers in pipes. This dragged air is called *entrained air* and helps mix up the supply air. If air is not entrained, it is called *stagnant air*.

Coanda Effect

Air tends to flow along surfaces rather than as a free jet in the air. This is called the *Coanda effect*. It is produced because air will take the path of least resistance, and there is less drag between a wall and an airstream than there is between an airstream and stagnant air. Part of this behavior is due to the creation of a low-pressure pocket between the airstream and the wall, which draws the airstream into the wall. The Coanda effect allows supply air to flow across the ceiling of a room, which can be used beneficially to push cool supply air far from the diffuser. Figure 9.3 illustrates how the discharge of a linear slot diffuser hugs the ceiling for a while before dropping into the occupied zone and mixing.

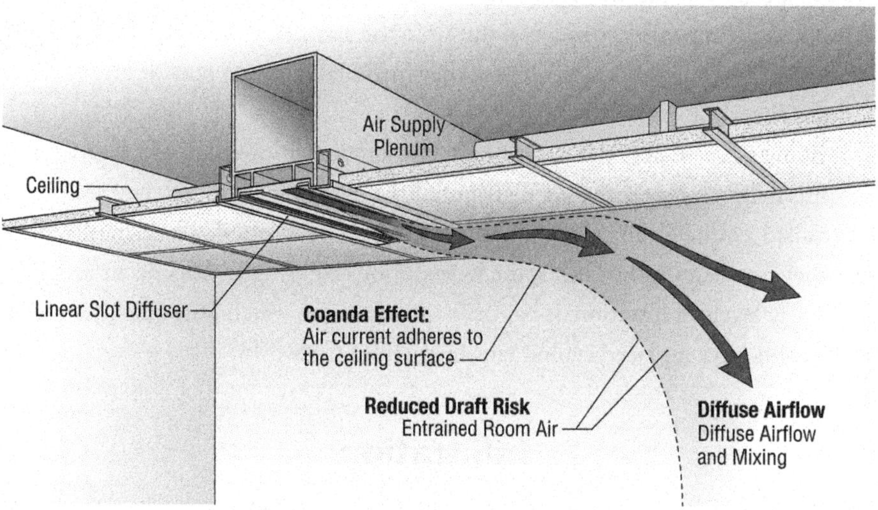

FIGURE 9.3 The Coanda effect.

Balancing

In constant air volume systems, pressure control is used to your advantage in ensuring air goes where you want it to. Pressure can be controlled by dampers, but as mentioned before, in the best constant air volume designs the pressures are designed to be equal by sizing the ducts carefully. It wastes money and material to have a huge duct that just has to be throttled back with a damper. It's cheaper to make the duct smaller and build the drag into the system where it is needed. Test, Adjusting, and Balancing (TAB) is the process of ensuring design optimization and is an important step in commissioning a system.

Non-Ducted Systems

We have been looking at forced air systems; however, there are other important systems that operate differently. Some systems, especially for residential and retrofit applications, send thermal energy with water pipes or refrigerant lines. Examples of this include systems such as radiant floor, baseboard, PTACs (packaged terminal air conditioners), and ductless mini-splits. These systems have the advantage of being quieter and using less space for ducts. These attributes make them especially attractive for retrofits. However, these designs have the disadvantage of not being able to condition the air in the same way as air handlers; namely, most don't filter or dehumidify it, and they can't blend it with outside air to reduce carbon dioxide levels.

Some of these non-ducted systems, like ductless mini-splits and PTACs, rely on internal fans to get the air distributed through the room, so they are not as controlled as forced-air systems. Mini-splits generally cool down recirculated air, but they can dehumidify and some have options to bring in outside air and provide basic levels of filtration. It has been interesting to watch how quickly ductless mini-splits have replaced ceiling fans in much of the world.

Infiltration

A poor cousin to air distribution is the effect of infiltration on buildings and houses. *Infiltration* is the word most often seen in thermal load analysis to cover uncontrolled (as opposed to controlled outside and exhaust air) heat exchange by mass transport. Infiltration is one of the building thermal loads that you use to size an HVAC system and try to reduce in the building design.

Infiltration is caused by pressure difference, which can normally be attributed to two things: (1) stack effect and (2) wind. The *stack effect* is caused when cold air pushes up hot air and produces strong convective cycles. This makes the bottom of the building act like a vacuum cleaner and draw in unconditioned, outside air. Wind can also produce infiltration. Understanding how this works is connected to our earlier discussion of pressure drag. Wind creates high pressure on the part of the building it smacks into. Wind also created low-pressure eddies on the leeward side of the building. These eddies act like a vacuum cleaner, drawing in unconditioned, outside air. This effect is reduced with infiltration barriers, from Tyvek to the newspapers I've seen taped to a log cabin I used in Kentucky. These sealed building envelopes make buildings more energy efficient but rely on the air handler for bringing in outside, fresh air.

SUMMARY OF WORRIES

Here are some things you should always be worrying about regarding air distribution:

1. They get blamed for everything.
2. They have to quietly mix air throughout the occupied zone.
3. More diffusers make ductwork design expensive. See number 2 above.
4. Diffusers will be used to hang decorations.
5. Return grilles will get covered by furniture.
6. Putting supply diffusers close to a return grille short-circuits the system and makes someone look very foolish.

TEN

INDOOR AIR QUALITY

When you consider pushing air into a room, you should also think about indoor air quality (IAQ). One of the biggest problems with indoor air quality is stale air—that is, the presence of too much carbon dioxide. This problem is managed by bringing in fresh, outside air through the air handler and mixing it with the supply air. In some designs, 100% of the supply air comes from outside air. This means that the return air is ducted entirely outside and new, fresh air is brought into the air handler. This outside air runs through the heating/cooling coils. This approach assures a low level of carbon dioxide in the supply air but decreases energy efficiency because all the return is exhausted outside. Typically, this exhaust air is run through a heat exchanger to extract some of the energy. This is done with a heat recovery ventilator (HRV) or an energy recovery ventilator (ERV).

Moisture

Humidity control is one of the biggest concerns in designing and installing a healthy HVAC system. Commercial systems are usually designed around cooling capacity because most buildings are self-heating and do not need much heating capability. However, moisture causes all sorts of problems, from corrosion to mold. Makeup air units can reduce the moisture of outside air before it's brought into the building. When moisture is removed with cooling coils in the air handler, the velocity of the supply air must be slow enough and the depth of the coil deep enough that there is sufficient time for the water to condense on the coils and drop out of the air. The condensate has to be routed outside, and any condensate pans have to have detection capability to make sure they don't overflow if the drainage system gets plugged or if the lift pump doesn't work. It is surprisingly problematic to remove moisture from air and send it to an outside drain. In some ways we struggle to take that too seriously because we're focused on conditioning the air. However, a well-built condensate system and controls are important for a healthy cooling system.

Pollutants

Pollutants can vary from irritants to things that cause an allergic reaction or even a disease. Diseases can be caused by a *pathogen*, which is a term for any organism that causes a disease. Pathogens differ from *allergens*, which produce an allergic reaction, because pathogens can bypass your immune system and multiply inside you.

Pollutants include gases like carbon monoxide, formaldehyde, and volatile organic compounds (VOCs). Health problems can also arise from solids such as the range of microbials that include viruses, fungi, mold, bacteria, nematodes, amoeba, pollen, dander, and mites.

Dangerous Bacteria, Fungi, and Gases

The most famous problems in the HVAC world are probably the bacterium *Legionella*, the fungus *Stachybotrys chartarum* (black mold), and the gases *radon* and *carbon monoxide*. In Chapter 4, "Moving Heat and Cold," we considered how biocides can be added to water systems to attack disease-causing pathogens. They usually work by destroying cell walls and internal cell structures. *Legionella* causes Legionnaires' disease, a type of lung infection. *Legionella* needs to have the protective biofilm stripped away before it can be attacked.

Large exposures to the fungus *Stachybotrys chartarum* can be dangerous. The fungus kingdom of organisms is the category that includes molds, yeasts, and mushrooms. (If you thought there are only two kingdoms, plants and animals, you missed the news that fungi have been separated from the plant kingdom since 1969.) Fungi are made of thread-like filaments called *hyphae* and reproduce through airborne spores. These spores are super strong and durable. They can be very hard to destroy.

Stachybotrys chartarum is a greenish-black color and may appear to have a wet appearance but is powdery when it dries. This mold will have a strong musty smell. However, it can only be positively identified by a microscopic exam, as shown in Figure 10.1, or by a polymerase chain reaction (PCR) test. Let's face it—everything moldy kind of looks like black mold.

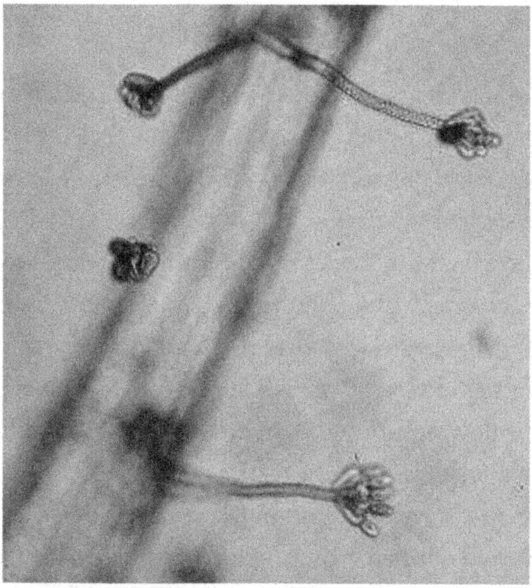

FIGURE 10.1 *Stachybotrys chartarum.* (*Source:* This image was created by user Joshua Laurila (joshlaurila-uwl) at Mushroom Observer, a source for mycological images. Wikimedia.)

Other common molds include:

- *Aspergillus* (yellow or bluish-green color)
- *Cladosporium* (black and green with a granular texture)
- *Penicillium* (blue, green, or yellow)
- Mildew, which is lighter colored than molds and is only a powdery fungal surface growth (commonly white or gray)

Some molds release chemicals, called *mycotoxins*, which can also be hazardous. The metabolism of fungi and bacteria can release *microbial volatile organic compounds* (MVOCs). *Endotoxins* are toxic chemicals that are released by bacteria when they disintegrate.

Radon is a concern in many areas because it can be related to lung cancer. It comes from some soils and rocks near natural uranium deposits. Like carbon monoxide, it is invisible, odorless, and tasteless. Radon is usually controlled by sealing and/or sub-slab depressurization to route the radon around living areas.

Air Filtration

In Chapter 2, "Making Heat," we looked at particulates and their effects on those who breathe them in. While pollution control of the boiler stack can involve baghouse filters, electrostatic precipitators (ESPs), and other advanced technologies, in this chapter we are looking only at indoor pollution, including odor control.

Indoor pollution is commonly reduced by using filters. Filtering fine particles from the air takes advantage of the molecular surface forces we talked about when we looked at phases. These surfaces force particles to adsorb (not absorb) to other things, such as air filter media. *Adsorb* is a fancy word for sticking.

Particles in air flowing through a filter will stick to the filter media and get removed from the airstream. This *interception* and *diffusion* effect causes the particles to be filtered in a way you might not expect. As shown in Figure 10.2, particles don't get strained out like water through a colander, but rather the particles are attracted to, and stick on, the air filter media. The medium is often a glass, like borosilicate, which can easily develop a static electric charge.

The erratic *Brownian motion* of very fine particles moves them around in many random directions, so they are more likely to collide with filter media than larger particles with a lot of inertia. Those big guys just want to fly straight through. These big particles are filtered in a different way and usually beforehand. *Electrostatic* filters induce a charge on the individual particles and use the opposite charge to draw them out of the airstream.

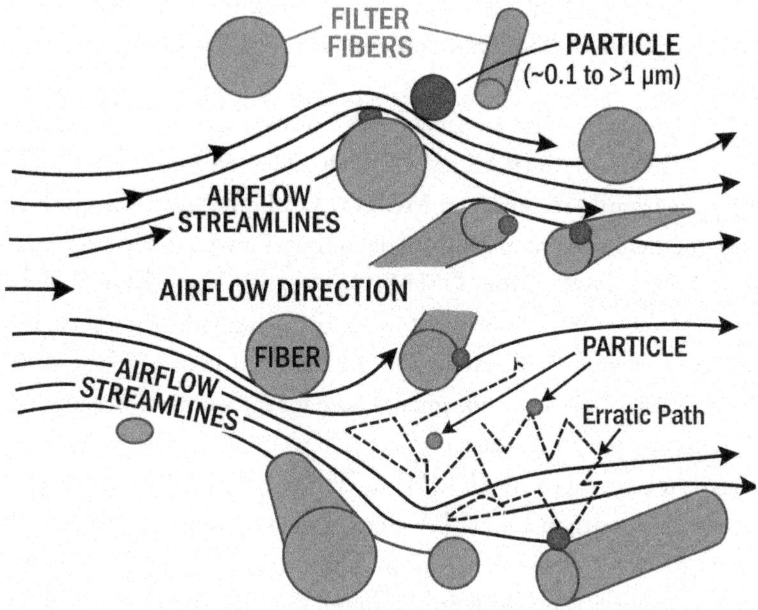

FIGURE 10.2 Interception and diffusion effect.

Air Filter Performance

Air filter performance is measured with MERV (minimum efficiency reporting values) ratings. The filters are rated by how small a particle they can remove. The common dimension for evaluation is the micron, which is 1 millionth of a meter, or 0.00004 inch. Human hair is approximately 70 microns, while the smallest thing you can see is about 40 microns. A human red blood cell is about 5 microns in diameter.

The MERV ratings go from 1 to 20, with 20 being the finest filtration. MERV 1 through MERV 4 filter out particles down to 10 microns, like pollen. MERV 5 through MERV 8 filter from 3 to 10 microns and are used to block things like dust and spores. Filters capture even smaller particles as you move to MERV 9 through MERV 12, which filter from 1 to 3 microns and are used to filter fine dust and *Legionella*. MERV 13 through MERV 16 filter from 0.3 to 1 micron. MERV

17 through MERV 20 filter less than 0.3 microns. The filters with the highest MERV values are used in clean rooms and operating rooms along with pressurization control.

HEPA (high efficiency particulate air) filters are very effective and are rated at around MERV 17. ASHRAE and the CDC (Centers for Disease Control and Prevention) currently recommend MERV 13 or higher filtration for buildings. Interestingly, there are a lot of different definitions for what HEPA stands for. The definition I used above is from OSHA. OSHA defines a HEPA filter as capable of being "at least 99.97% efficient in removing monodisperse particles of 0.3 micrometers in diameter." Reference OSHA Standard 1926.1153 for more information. This is also the source document I used for the words corresponding to the HEPA acronym (people disagree on this . . .).

Ultra-low particulate air (ULPA) filters are even more effective at removing tiny particulates than are HEPA filters. This filtration improvement is made by packing the filter media more tightly. However, nothing is free. As filtration performance increases, pressure drop through the filter increases. Therefore, you need bigger fans that consume more electricity. Cleaner air means higher adverse environmental impact. Balancing cleanliness and environmental impact, as well as capital cost and operating costs, is just one of the many judgments required in designing systems.

Embodied Energy

It is an interesting aside to note the relevance of cradle-to-grave embodied energy when considering environmental impact. This notion considers the power and pollution associated with things that are often not thought about, such as mining, processing, transportation, packaging, and managing materials and products. Life cycle assessment (LCA) is one of the management tools used to analyze and present this sort of information. It is defined by the ISO standards 14040 and 14044.

How much interior air filtering should we have? Is filtering interior air good for the environment? The answers are not easy in the context of life cycle assessment.

UV Light and Pressurization

Some HVAC systems use ultraviolet (UV-C) light to kill pathogens in the air supply. The energy in electromagnetic radiation is a function of its frequency. The higher the frequency, the more energy it has. Therefore, UV light has more energy than visible light. While infrared energy feels warm to the skin, and visible light is what we see, UV light has the highest energy of these three bands of electromagnetic energy and is best for killing bad actors in the air supply.

ASHRAE Standards 62.1 and 62.2

Indoor air quality is an important issue; however, we must understand that these chemicals aren't like poison. Not all black-colored mold is *Stachybotrys*. The health effect of these pollutants is usually related to the quantity and duration of exposure. Fortunately, there are standards upon which to lean in this nebulous topic of indoor air quality. Refer to ASHRAE Standard 62.1 (Ventilation for Acceptable Indoor Air Quality) and Standard 62.2 (Ventilation for Acceptable Indoor Air Quality in Residential Buildings) for information on specific recommendations on this topic.

SUMMARY OF WORRIES

Here are some things you should always be worrying about regarding indoor air quality:

1. All mold looks like black mold.
2. Everyone thinks they deserve the best air filtration possible.
3. Everyone wants 100% outside air.
4. Pressure control can be accidentally modified.

ELEVEN

CORNY STORY

Corny Story 4

Journey from the Air Handler

A final corny story about a day in the life of an air handler.

You're outside on a hot and humid day. You're the oxygen molecule, and your family of four nitrogen molecules are with you. You were just hanging around outside when suddenly you got drawn into an air handler. A fan pulls you in, and another one pushes you through an air filter and coils. The coil is cold, and you can feel your heat leaving as you slow down your happy, summer-day bouncy motion. You huddle closer to your family, feeling your density increase but enjoying the coziness.

You look over your molecular shoulder and see some water molecules have joined you from outside. A creepy smile forms on your face as you watch them starting to stick to each other in the cooling coil. At first, they would hit each other and stick for a while and then shake free, but as they got colder and colder, their thermal motion wasn't enough to free them from the grasp of each other's surface forces (what happened to the hand magnets?). They became drips and fell to the bottom. "Losers!" you call out, as you watch them go from gas to a liquid phase.

Next thing you know, you're running through a silver duct. You fly through a fire and smoke damper and around corners until you start getting pushed through a round duct coming out the side of your original duct. "That was easy," you say to your neighbors as a big metal turning vane guides you smoothly to your new direction. You all moved together, never changing your speed. You slide into the smaller duct. There are fewer of you now.

You travel a little while and then are pushed into some weird diffuser thing with metal slats in it. Suddenly, you are flying down into a room full of happy people. Your velocity slows down, and soon you and your family are plucking the back of the neck of some guy drinking coffee. He says, "It's really comfortable here. It's quiet, and I can barely feel the air conditioning on my neck."

You linger around in the room for a while as you slowly heat up. Eventually, you feel you're being pulled to a wall. As you get closer to the wall, you see there's a metal grille discretely located in the wall. It's painted the same yucky cream color as the wall. You and your family get sucked into the wall-colored grille and the dusty duct attached to it. You're moving back to where you came from, but this time you're in a bigger duct and moving a little slower. You notice a few carbon dioxide molecules heading back with you as you go back to the air handler that first sucked you in from outside. Next thing you know is that you're spit back outside. "That was fun, I guess," you tell your family as you float up, enjoying the freedom of the open sky.

On another day someone closed the outside air damper. However, as it turns out through the miracle of corny storytelling, you're already in the air handler—you and your family of four nitrogen molecules. You are still the glamorous oxygen molecule, not that it matters because no one's trying to burn anything.

You get pushed through the filter and a coil again. Then you rush down the same silver duct until you're back into the same room and plunking the back of the neck of someone. You get drawn back into the return grille on the wall. You're joined by some carbon dioxide molecules. Instead of being exhausted outside, like you were last time, you're recirculated back into the air handler. You complete the cycle again and again and again, just like you did in the first part of this corny story. However, now you notice there are more and more carbon dioxide molecules each time.

"Those guys in there must be breathing a lot," you explain to your friends.

"He's such a know-it-all," they whisper to each other.

Then one of them shouts out, "I heard they're installing an ERV and we'll all be dumped outside soon."

"That'll be nice. I like sunlight," you reply, thinking about how cool you look in molecular shades.

SECTION 3

EVERYTHING ELSE

This title is a little presumptuous, but let's look at a variety of things. Bored with a topic? Turn the page.

TWELVE

NATURE

Nature challenges a lot of HVAC equipment. These challenges range from bugs getting into the fill of cooling towers to lightning strikes traveling through corrugated stainless steel tubing (CSST). Vermin eating your wire insulation is just one of the many threats to systems exposed to nature. You are always trying to keep water and bugs out of your equipment. You are also thinking about lightning strikes, snow, freezing rain, salt spray, and all the weird stuff you have in your area. It's rough out there.

Weather

The biggest concern with weather is the temperature and humidity with which your system will have to contend. These values are the first thing entered into thermal load calculations for buildings. Outside temperature and humidity affect not only a building's thermal load, but the amount of conditioning required on the outside air intake, efficiency of evaporation in cooling towers, efficiency of electric motors, performance of lubricants, etc. Weather also produces lightning, hail, snow, rain, wind, pressure, and lots of peculiar systems from killer monsoons and tornadoes to dust devils and sneaky cold katabatic winds that slide down mountains.

Atmospheric temperature is created by the sun and varies as a function of air movement. Large-scale, global winds are motivated by the uneven heating of the earth between the hot equator and the cold poles. Temperatures are also regulated by contact with land and sea.

Wind is produced as air moves from high to low pressure. Monsoon winds are driven by the different responses to heating between land and water. For example, the winds in southern Asia are largely driven by the sinking of frigid air in landlocked central Asia that pushes outwardly. The winds turn around when the air in central Asia heats up, causing the air to rise. This rising air sucks in the air from southern Asia.

Low-velocity "wind" is created in buildings by pressure difference also. This happens when wind produces high pressure when it smacks against a building, while on the lee side of the building, eddies often occur that produce low pressure. This difference in pressure causes uncontrolled infiltration of air into and exfiltration of air out of the building. While we like the fresh, low carbon dioxide air coming into a building, we don't like uncontrolled air that has to be heated or cooled and dehumidified.

Wind also carries pollutants, ranging from tiny solid particulates (that get deeply inhaled into the lungs) to corrosive chemicals. Outdoor particulates and other pollutants get into buildings through the outside air intake and uncontrolled infiltration. The density of these pollutants can vary with wind velocity (or lack of it) and direction. The production of particulates and indoor air quality filtration will be discussed later. However, we should recognize that outside air can contain nasty stuff that varies with wind behavior.

Wind and pressure can affect the behavior of some venting systems by pushing air down into the stack. This downdrafting can result in several types of problems, from odor to dangerous lack of combustion venting. Cold air above a stack or vent will tend to sink and produce a downdraft. Wind speed, atmospheric pressure, and humidity also affect draft.

Natural Threats

Nature is attacking your equipment. Bugs get everywhere, and their webs, nests, cocoons, body parts, and all the other yucky stuff plug things. Bigger animals build nests that can block vents and intake grilles, and vermin chew through electrical wires and make nests out of insulation. Rain doesn't always go where it was designed to flow. The splashing, dripping, and streaming of water will find nooks and crannies (and circuit boards). Sealing systems are imperfect, and expanding foam is not a cure for everything. Water expands as it freezes and breaks stuff and opens up cracks. Dust slips through holes even too small for a mouse to enter, and a piece of plastic garbage can seal a pipe. I'll stop there, but this is an ongoing list, and people work really hard to keep nature away from equipment. I shouldn't be so cynical. Often sealing systems are good, but so many get compromised over time. I have too many stories, from discovering heated pump stations with hundreds of mice living inside to sticking my hand in a pile of decaying mice carcasses. I'll stop there, before I start talking about bugs.

Light

From our HVAC standpoint, we know that light sources produce heat. This heat production is much less now with modern lighting technology and "daylighting" approaches to building design, but it is part of the environment that makes commercial buildings self-heating. Outdoor light is a different matter. Visible light lets us see the world around us, which is nice. But light has many more things to consider than just illumination.

The energy of electromagnetic radiation is directly related to its frequency. This mathematical relationship is described in *Planck's law* as radiation energy is equal to its frequency times a constant (called *Planck's constant*, which seems logical). Low frequencies, like AM radio, have very little energy, while high frequencies, like X-rays, have lots of energy. Between these extremes are infrared, visible, and ultraviolet light. Infrared light is *light* (electromagnetic radiation) wavelengths just below visible light. After visible light, this is the frequency spectrum that you are most familiar with because it produces the heating effect on your skin. We use this experience advantageously for heating areas with high air movement, such as vestibules and outdoor seating areas. Figure 12.1 shows the electromagnetic spectrum relating frequency, wavelength, and common nomenclature.

Not only do electromagnetic waves have different frequencies (and corresponding wavelengths), but the physical length of the beam is a real quantity. That is, electromagnetic waves have a beginning and end, just like a wiggly caterpillar. The single piece of an electromagnetic wave is called a *photon*. The mixture of an electromagnetic wave and physical length is what gives radiation both wave and particle properties.

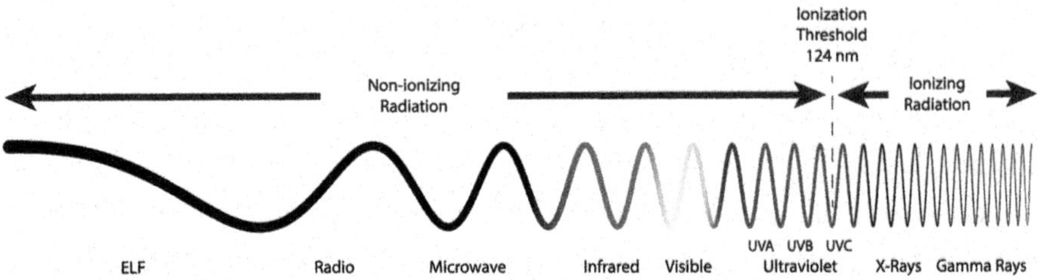

FIGURE 12.1 Electromagnetic spectrum relating frequency, wavelength, and common nomenclature. (*Source: Lashb007, CC BY-SA 4.0 https://creativecommons.org/licenses/by-sa/4.0, via Wikimedia Commons.*)

Electromagnetic radiation frequencies just above visible light are called *ultraviolet*. This radiation can cause *ionization* (dislodging an electron from a molecule), break chemical bonds, and damage DNA in living cells. This radiation has more energy and shorter wavelengths than visible light. Half the UV light reaching us doesn't come directly from the sun; it comes from the UV bouncing around in the sky. This bouncing behavior makes it difficult to block UV light with physical things like a shade.

Ultraviolet light (or electromagnetic radiation) is the high-energy, high-frequency part of light sources such as the sun. The ultraviolet light spectrum is further broken into three categories: A, B, and C. The wavelength for ultraviolet light is measured in angstroms; an angstrom is 100 millionth of a centimeter. T he longest wavelength of UV light, referred to as *UV-A*, has a wavelength in the range of 3,200 to 4,000 angstroms. Longer wavelengths equate to lower frequencies. The UV-A spectrum penetrates the atmosphere. UV-B light is the next highest frequency/short wavelength range, and it is described as electromagnetic radiation within the frequencies of 2,800 to 3,200 angstroms. This spectrum is even nastier for human health. This frequency range is filtered by the earth's ozone layer. Ozone is a molecule made of a fairly unstable assembly of three oxygen atoms. It is most famously known in the HVAC world as something that's easily destroyed by chlorofluorocarbons (CFC) and halon. This ozone damaging effect was one of the earliest motivators for phasing out the use of refrigerants such as CFC.

Windows are very good at absorbing UV light. Even a single pane of glass blocks almost all UV-B light and more than half of the UV-A spectrum. This UV light penetration is what causes fabrics to fade. Plants are good for a lot of reasons. One of them is because green plants absorb UV very well—less than 10% of the UV light reflects off plants and back into the environment. Because UV light is scattered by the atmosphere, the more atmosphere the sun has to penetrate, the less you receive. Mornings and evenings in the winter at sea level are the best times and locations for minimizing UV radiation.

UV-C spectrum radiation is deadly. It has a very high frequency, high energy, and a short wavelength (2,000 to 2,800 angstroms or 200 to 280 nm). It is very good at killing things and is used in water treatment and other processes. Fortunately, UV-C does not penetrate our atmosphere. However, when it is released by UV light disinfection systems, it can ravage mold, viruses, and bacteria by disrupting their DNA. We will also consider low-E windows and solar energy applications in Chapter 14, "Alternative Energy."

THIRTEEN

ELECTRONICS: BUILDING AUTOMATION, COMMUNICATION, AND MOTORS

Building Automation

Connectivity of equipment and the sharing of data are vital to modern HVAC operation. All your pieces of equipment are talking to each other. Then they collect data and use the data to anticipate how the equipment should operate. Mass amounts of data are being passed around between equipment, controls, and large networks. The data goes well past being measured in gigabytes, quickly cascading into terabytes, petabytes, and zettabytes (1 zettabyte equals 1 trillion gigabytes). This science fiction soundtrack just shows how much data is being pushed around, analyzed, and acted upon. Sometimes there are tiny delays in communicating data, called *latency*. This delay causes everyone to get frustrated and offers a great generic excuse for network problems. It's sort of "the dog ate my homework" of the IT world.

Automation is changing fast. HVAC controls are getting smarter, the amount of data gets bigger, and the ability to predict and diagnose problems escalates quickly. This interconnectivity and accumulation of data allows building automation systems to improve HVAC efficiency, which is a good thing. A lot of this improvement is based on AI-developed rules that are created as the systems operate. AI lets the controls network learn how equipment is behaving as well as the patterns of occupancy and equipment usage. AI develops rules that it expects will work in the future based on information it has already acquired.

We're a lot smarter than computers—we use sophisticated and intricate thought processes to solve problems, recall information, and make decisions. Acquiring knowledge and creating rules like a human expert is tough. However, computers are excellent at acquiring huge amounts of data, and AI can use it with a peculiar flavor of human intelligence.

The automation of buildings allows for centralized communications and controls. These systems are often called *building automation systems* (BAS), but there are other names and acronyms that are also used. These centralized systems allow the monitoring and control of building infrastructure such as HVAC, lighting, communications, security, and fire systems. A BAS uses a network of sensors that detect a wide array of information. Some of this information might include the status of the temperature, humidity, carbon dioxide, occupancy, lighting, door openings, and equipment function. The sensory data is routed to a control system, which in turn has outputs that allow the control of equipment and communications. Not only do these systems allow higher energy efficiency, but they can improve comfort, safety, and general waste of resources. In addition, they can track utility consumption such as electricity, gas, and water. BAS also manage security systems that can include smoke and fire control and alarming.

A BAS uses standard digital language to send commands and receive data. These "languages" are standard industry protocols, and there are so many protocols that I am reluctant to list them. Some protocols, such as BACnet, are part of national standards (in this case, ANSI/ASHRAE Standard 135), and some are proprietary. We are not going to look at these protocols or the interaction designs, but we will look at the basics of electricity and wireless communication.

Controls

Control systems have moved from my world of programmable logic controllers (PLC) to click-and-drag, networked systems. Control systems send signals either on wires or with wireless electromagnetic radiation. For hardwired connections, the information is sent as flowing electrons in a conductor, namely current through a wire. Electron flow needs to overcome the electrical resistance to flow, which is why we use highly conductive wire, like copper. The electrical resistance is overcome with a voltage source, called an *electromotive force*; that's why the letter *E* is used in electrical calculations. Power is equal to the voltage multiplied by the current, and the power loss in wire is the product of the current squared times the resistance. This is used to your advantage if you're trying to heat something, but commonly it's a disadvantage because it wastes energy. If you're sending signals a long distance, you want high voltage because you can use lower current. Power loss is a function of the current squared. If you double the current flow, you create four times the power loss. Wire sizing is dictated by most codes to limit the voltage drop in a system. Low-voltage systems are attractive in many applications, especially controls, because they avoid the electrical codes that strictly govern high-voltage systems.

Communication

Data can also be sent by wireless electromagnetic radiation like the radio waves you pick up on your car radio. Electromagnetic radiation has too many syllables, so I'll just call it *radio signals*.

Radio signals are sent by either turning the radio signal on and off or by imposing information on top of the radio signal. The receiver separates the information from the carrier frequency. The frequencies used in control systems are high enough that they don't get bent or distorted very much, whereas low-frequency signals can get readily bent by the ionosphere (a layer of high-altitude, ionized air) and reflect back to earth.

Matching the transmitter or receiver to the transmission line and antenna is done by designing these systems so their impedances are the same. *Impedance* is a measure of resistance to alternating current. It includes electrical resistance and electromagnetic inductive and capacitive losses. Inductive losses occur with motor windings and other wound wire coils, while capacitive losses occur with things that, well, are capacitive . . . We'll talk more about this in the Appendix and see how it relates to power and efficiency.

Because inductive and capacitive losses vary with frequency, the impedance value also depends on frequency. There is one "tuned" frequency in which both the inductive and capacitive resistance (called *reactance*) are at a minimum. By changing the inductive or capacitive value, you can change the tuned frequency. Changing reactance is also used to purposely block frequencies.

An abrupt change in impedance, say between a transmission line and antenna, will cause the signal to reflect. Therefore, communication systems ensure that the impedance of all the devices matches. Impedance matching can be kind of thought of with a mechanical analog. It's sort of like transitioning from a fan discharge into a duct plenum.

Because antenna impedance is a function of frequency, the size of the transmitting and receiving antennas is dependent on the frequency they are intended for. High frequencies require shorter antennas than do low frequencies. Radio signals used in control systems use high-frequency, short-wavelength signals in the ultra high frequency (*UHF*) or super high frequency (*SHF*) bands. Therefore, the antennas are tiny. These high frequencies allow a lot of information to be transmitted. However, these signals can be stopped by a sheathing of conductive material. This conductive sheathing is called a *Faraday shield* (or sometimes a *Faraday cage*). The sheathing is used for setting up electronic equipment in an environment where it won't be exposed to outside radio waves. When you go in one, it's funny to watch how your phone loses any kind of signal strength. Sometimes Faraday shields are inadvertent; one project I worked on had a copper roof, and it acted much like a Faraday shield and blocked radio signals.

A cousin of the Faraday shield is a *grounding system*. It ensures that there is a shared electrical ground for all the equipment. When this is not the case, noise can enter the wires. This problem of slightly different grounds is called a *ground loop*. Ground loops are eliminated by wiring practices that ensure all the grounds for equipment are electrically connected to a single ground point. Our dear friend and noisy companion, the National Electric Code, has a lot of detail on grounding and bonding systems. Go digest the NEC if you want to learn more about best practices in this area.

Motors

Electrons flowing in a wire create a magnetic field. Creating and moving magnetic fields is used to great advantage in electric motors by allowing two magnetic sources in the motor to interact with each other. This interaction is designed so the stator magnetic field and rotor magnetic field develop an attraction and repulsion that causes rotation. The continuous magnetic pushing on the rotor is done by commutation. *Commutation* is the fancy term for switching the current direction in the motor and therefore changing the magnetic field that spins the rotor. Motors have typically used contacting brushes that sweep along the rotor (actually the rotor's commutator) and switch the electric current. Brushless motors replace this contacting system with noncontacting sensors that indicate the rotor position and then electronically switch current to different stator windings (see Figure 13.1). Brushless motors have higher efficiency and require less maintenance. They are so much in vogue now that they have their own special acronym: BLAC for brushless AC motors and BLDC for brushless DC motors.

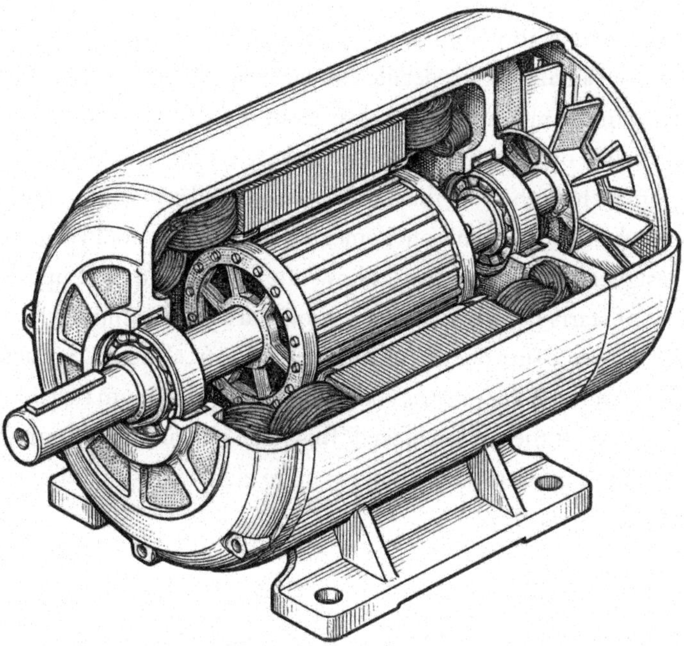

FIGURE 13.1 Induction motor.

There are a variety of motors used in HVAC. The king of the heap is the efficient, reliable three-phase motor that runs big fans and pumps. This kind of motor is exquisitely made and runs reliably for a long time. A *permanent magnet synchronous motor* (PMSM) is another high-efficiency AC motor. It uses permanent magnets in the rotor and can produce high torques at low speeds. Smaller stuff is often handled by an *electronically commutated motor* (ECM), which is a very efficient brushless DC motor.

The most practical consideration many people have with motors is changing the start capacitor on a single-phase motor. Capacitors store electric charge and can discharge it quickly. This flow of electricity gets a motor started so that the magnetic fields of the rotor and stator can interact with each other. This is often the first thing technicians check when there is a motor problem. You may find some old clocks in which you spin a knob to get the electric motor started; those were certainly simple and reliable ways to start a motor turning.

Inrush Current

Another vital consideration for some big motors is the maximum starts per hour. I have always been amazed at the detailed understanding that electrical fuse designers and others in the motor industry have about the first microseconds of current after it is switched on. This first flow of electricity is called the *inrush current*, and it is an enormously high value that quickly heats up the motor windings. If fuse designers didn't understand this phenomenon, all our fuses would be tripping, and controls designers would not restrict how often you could start up big motors. The number of starts per hour is usually provided by the manufacturer. NEMA, our friends that make sure our electrical panels are safe, also has standards on this. NEMA also governs *locked rotor current*, which is the maximum sustained current draw from a motor if the rotor is locked. This value is vital for wire sizing and fusing because this condition creates current up to eight times the full load current that heats things up fast.

Variable-Frequency Drives

Variable-frequency drives (VFD) have allowed wonderful improvement in motor efficiency. I have heard more acronyms and slang terms for VFDs than for any other components I can think of. The quirkiest, but highly descriptive, is *freq. drive*, pronounced "freak drive."

Running a motor is a huge cost in HVAC systems, especially commercial systems. Anything done to improve motor efficiency has a rapid return on investment. VFDs are one of the stars. They keep improving every year, and it has been a delight to watch them get used in more and more applications.

VFDs work by changing the frequency of the AC driving a motor, which in turn changes the speed of the motor. VFDs can be networked so feedback from actual operating conditions, like pressure and temperature, can be processed and fed back to the VFD to speed up or slow down motors.

SUMMARY OF WORRIES

Here are some things you should always be worrying about regarding controls and motors:

1. Control problems are the easiest and cleanest problem to fix.
2. Controls are the most common problem in HVAC systems. Yes, the problem is usually the sensors, relays, or other electrical items, not the control boards themselves.
3. Controls do not cut energy costs as much as salespeople promise. The controls always struggle to live up to outlandish promises made by others.
4. Other things in motors can fail besides the start-up capacitors.
5. Motors that start and stop frequently heat up. They can be designed for a high number of starts, but they don't have to be.

FOURTEEN

ALTERNATIVE ENERGY

Wind and hydropower can be understood using fluid dynamics and thermodynamic theory. Solar energy can be described using heat transfer theory, specifically radiant heat transfer. However, some forms of solar energy rely on different theories. For example, neither direct electrical production nor cooling generation from solid-state devices is explained by the theories previously discussed. So let's do an overview of solar, wind, and hydropower in the context of what we have already learned. Then we can look at some of the theories behind solid-state electrical production (photovoltaic effect) and coldness generation (Peltier effect).

You can review the discussion of life cycle assessment and embodied energy in Chapter 10, "Indoor Air Quality," to give some context for this discussion.

Wind and Hydro

Wind-generated power and water-generated power can each have different names. Wind-generated power can be called *wind power* or *wind energy*. Water-generated power can be called *hydropower* or *hydroelectric power*.

Devices that use fluids to turn turbines normally use airfoil-type blades that produce a low pressure on one side and a high pressure on the other side. Wind turbine blades are an example of this feature. The turbine blades are designed so the pressure difference that rotates the blade is produced by forcing air at different velocities. On the "top" surface of the airfoil, the air has to travel farther and therefore faster. According to the first law of thermodynamics, and the Bernoulli equation that is derived from it, this high velocity forces the pressure to be lower. The pressure difference produces thrust.

Changing energy forms from velocity to pressure is also used in hydroelectric plants, such as using long, skinny tubes with turbines at the end or by using impulse turbines. Skinny tubes, called *penstocks*, accelerate water before it reaches a turbine that is rotated by the fast water. *Impulse turbines* rely on the velocity of water impacting a turbine blade and converting that kinetic energy into pressure on the turbine blade.

Solar

The sun emits electromagnetic waves at various frequencies. A lot of this radiation transmits through our atmosphere and strikes the earth. We happen to see a small band of visible light, but frequencies much higher and lower than visible light make it through the ozone and other disruptive layers of our atmosphere.

Solar energy can be turned directly into electricity with photovoltaic cells. Solar energy can also be tapped by using thermal systems that capture the solar energy and store it. Buildings can be designed to stay cool and minimize unwanted solar heating, or in the case of homes in cold climates, can maximize solar heating in the winter. This is where architects and HVAC people need to work together.

We looked at radiation heat transfer previously, and all that theory applies to solar energy. It is worth noting that if you are trying to reduce heat gain on a building from the sun, you want to maximize the exposed material's reflectivity and minimize the absorption and transmission of infrared energy. This is done by making buildings opaque, so that they block sunlight, and using reflective surfaces like white or aluminized coatings. Obviously, glass roofs are not a good idea if you are trying to keep things cool! Conversely, if you are trying to improve solar heat gain, the glass roof might make sense. A black roof would also improve solar gain because it could soak up heat and get crazy hot in the sun. If you want to maximize solar heating, you paint surfaces black and let the sunlight get absorbed! Of course, that doesn't help at night and when the sun gets obscured by clouds.

Emissivity is the willingness of a material to radiate energy away from itself. If you are trying to get rid of heat, you want to maximize emissivity by doing things like painting surfaces black. That is why the SR71 Blackbird airplane (the now-retired spy plane) was painted black—it helped emit the heat built up during high-speed Mach 3 travel. There is no other way to get rid of heat when you're in whisper-thin atmosphere and there is nothing to conduct heat to. While the black color will absorb more solar radiation than another color, this is small compared with the heat that has to be removed. Of course, the black made it harder to see the plane, and the paint had ferrite particles that absorbed radar signals, which is a good thing for a spy plane. On the other hand, if you don't want to get rid of heat, use a low-emissivity material, like aluminum or chrome. The *E* in *low-E* windows stands for *emissivity*. These windows are coated with a thin, transparent layer of a material that does not emit radiant heat.

As it turns out, emissivity equals absorptivity. A perfect "black box" surface absorbs 100% of what strikes it and emits 100% of its internal heat, just like a perfect mirror reflects 100% of all the visible light that strikes it. In all these radiation cases, the heat transfer is still a function of area and temperature difference. More heat is absorbed and emitted when you have a large surface area and a big temperature difference.

You may hear the term *albedo* used in architectural discussions. This value is like reflectivity but refers specifically to the percentage of solar radiation reflected by a surface.

If you absorb heat efficiently, say with black, water-filled pipes, you can store the heated water and use it at night. All-air systems are also used in this way. Air can be heated during the day and directed to flow around rocks or things with a high thermal mass and a lot of surface area. When the heat is needed, you let cold air flow through the heated rocks (or whatever). The air picks up the heat and can be used to keep your toes warm. Using thermal mass as a storage device is an old trick.

Solid-State Marvels

Photovoltaics

Atoms have a cloud of electrons around their nucleus. If the electrons move from the atom's orbit, they produce an electric current. Photovoltaic (PV) cells are made with two different layers of silicon that have loosely bound electrons. When sunlight is absorbed by the cell, the sunlight has enough energy to kick electrons out of their orbit, and they become free to move through the material. This creates a DC electric current, just like a battery. As you might imagine, the process by which certain materials react to photons of light this way is called the *photovoltaic effect.*

Solid-State Cooling

While not an alternative energy device, it is helpful to look at thermoelectric coolers briefly because, like photovoltaics, they use only solid materials. These devices rely upon the *Peltier effect*, which is the production of cooling when a DC voltage is applied across two semiconductor devices of different electron density. That is, you get cooling produced directly by a set of materials that have a DC voltage applied across them. The amount of cooling depends on the amount of current. Certain materials, like silicon-germanium alloys, will respond to DC voltage by having one side cool down and the other heat up. Of course, heating things up with electricity is easy, and you might enjoy the glow of your toaster wires every morning. However, making coldness directly with electricity can be useful. While not powerful, thermoelectric coolers can produce cooling in precise locations, and there are no moving parts.

SUMMARY OF WORRIES

Here are some things you should always be worrying about regarding alternative energy:

1. Nature doesn't just want to give you energy; it wants to destroy your equipment.
2. A significant amount of embodied energy is consumed in manufacturing and installing these systems.
3. The systems can have low energy density and reliability.
4. There are unanticipated consequences.

FIFTEEN

CLOSING THOUGHTS

I have had two stages of work, one as a practicing engineer and designer, the other as a college instructor. In my professional career, I have designed all kinds of products and systems. I have served on technical committees and a company board. But when I started teaching at the age of forty, the things we designed and made got really weird. During that time, I worked with my students (and my own kids) to make wonderful things, like an optically-clear ice maker, wind tunnel, snow maker, two-stage refrigeration system, a jet engine–like thing, experimental burners, chillers, and even a fire-shooting glove. The foundational knowledge of thermal science made it fun to experiment with things.

The good news about investing time and energy into understanding the foundations of thermal sciences and the other disciplines that intermingle in HVAC is that the concepts don't change over time. They are steadfastly rooted in physics, and they don't change even when shiny new equipment is introduced and marketing driven terminology is lavished upon the latest trend. Fortunately, shiny new equipment backed by marketing departments is often more energy efficient and quieter than the dusty stuff it replaces. However, the ways to make heat and cold and turn energy into pressure and velocity don't change that much. Neither do ways to efficiently move energy around a building or transfer heat and cold from one fluid to another. Our body is full of pipes and pumps, and the mechanical world, in its most optimal form, is likely to stay on that trajectory also.

However, the world of controls and electronics is vastly different than the mechanical world. That world changes very quickly, and the old-fashioned ways of ladder logic and linear programming have been replaced by intuitive interaction designs that use gestural approaches and graphical interfaces. Building automation and controls in general have provided great advances in optimizing equipment capacity and efficiency as well as making buildings more safe and secure.

Thank you for joining me in this engineering discussion of HVAC. A lot of this material was challenging, and mixing in the fancy words made it even more difficult to read. Thanks for your patience. Every one of the engineering-specific terms can be fully replaced with a sentence of your own making. The words are a shortcut; they don't present anything you don't know.

What we discussed in this book was robust and perhaps ponderous to plow through, but I appreciate your patience. Because the role of HVAC is usually to move energy around and keep people safe, the immutable principles that you've invested time into studying will serve you well. Thanks for your work in the HVAC industry. You make our world comfortable, productive, and safe.

APPENDIX

SUMMARY OF HVAC RULES, USEFUL EQUATIONS, AND OTHER HELPFUL INFORMATION

Introduction

(Who writes an introduction to an Appendix?)

"Why should I even bother learning this stuff? Can't computers do all this for me?"

Yes indeed, and they do a great job.

Many decades ago, I designed a cold-weather environmental chamber. I carefully calculated the system requirements. My wise boss told me to add a huge safety factor to my calculations. I was dismayed. Why do all the calculations if you just throw in a giant safety factor? He, of course, was right. If it weren't for that safety factor, it would not have pulled down cold enough. The bottom line is that rules of thumb can do a great job of including the real-life variables that are hard to jam into numbers and equations. Now when I do calculations like this, I use a variety of approaches to ensure I have the right order of magnitude from my calculations, but I also look at what has worked successfully in the past. However, there is a place for understanding mathematical concepts and presentations. This is your profession after all, and you want to have deep insights into everything associated with it.

About 35 years ago, I learned to use a computer program called *BLAST*. This was a great acronym for the geeky-labeled program called *Building Load and System Thermodynamics*. I still have the little round pin the trainer awarded me for learning the program. This program could do incredible calculations that really amazed me.

And software has come a long way since.

However, manual calculations let you check the computer calculations. Engineers traditionally verify analyses explicitly to make sure computer programs are reasonably correct. There are some engineers and designers who are completely entrapped by computer programs, and you have to wonder if they really even understand what's going on. So in this Appendix, we'll take a look at some of the governing equations driving engineering analyses. However, this book is *not* about sizing or specifying equipment and systems. That is way beyond the scope of this writing. Nevertheless, what you have read so far and what lies ahead in this Appendix are the foundations for much of the engineering and design work in HVAC.

Before we look at these equations, I ask that you please don't be intimidated by them. They only give you relationships that are not open to debate. The equations are here for those that like this sort of thing and find them useful in gaining insights into relationships between variables. I generally don't include units, and where I do, they are English units. I don't propose that you make these calculations; they are presented to show relationships, not to calculate values.

I was educated with metric units but learned to think in English units once I had a real job. I'll discuss "English" units versus geekier *fps* and *ips* units later. For those that prefer metric, I will refer you to the internet.

With the equations that are presented here, just think about the mathematical relationships between the variables, not solving for a specific number. Equations can be a lot more precise and efficient than long narratives and hand waving. Or even corny stories . . .

Note that the Greek symbol Δ just means change. For example, ΔT means a change in temperature. Other symbols are explained after the equations; perhaps the most important to note is that lowercase q and uppercase Q both equal heat energy. Lowercase q is used for rate of heat transfer (such as BTUH or BTU/hr), while uppercase Q represents total heat energy, as in BTU. Also, to help readability, I put parentheses around some of the weird variable names that have multiple letters, like (ACH), which is air changes per hour. These are commonly used in HVAC but drive mathematicians crazy. They don't like using the term q instead of dq/dt either. I'll expand on the meaning of dq/dt and its cousin dx/dt in the next section.

A final, humbling note. Sizing HVAC system components is often based on velocity, which is not determined by some magical science. Recommended velocities are provided as a starting point for calculations. These recommended velocities can be based on economics, such as piping; noise and comfort, such as air distribution; and rules of thumb, such as face velocities for air handler coils, stack velocity, and pumps. Therefore, using some of these calculations can start with an informed guess.

First, a Word on Math

Math is a tool, like a shovel. It helps you understand things. But I realize it can also create stress for many people.

The biggest changes in the way your mind thinks about math were:

1. Learning to count. This gave you the idea of numbers having a logic of their own.
2. Learning that $1 + 1 = 2$. This introduced manipulation of numbers.
3. Learning that if $x + 1 = 2$, then $x = 1$. This introduced abstraction of ideas into mathematics.
4. Learning that a change in something can be shown mathematically. For example, traveling a certain distance in a certain time lets you calculate velocity.

Let's expand this a bit and look at how you calculate velocity. You can consider the change in a variable and consider it over a tiny range—for example, the change in distance divided by the change in time gives velocity. But these changes can be looked at in an infinitesimally small gap in time so you can calculate the velocity at any point in time. This could be shown with the term dx/dt, or with an x with a dot or a tick mark over it. The *dx*/*dt* is a way of showing change over time and is the same presentation I used for the rate of heat transfer (*dq*/*dt*). Partial differentials consider cases with multiple independently changing variables. They are written with the fancy Greek letter ∂ instead of just a boring *d*. For example, if you are only looking at how the variable *x* changes with respect to time, it is shown as $\partial x/\partial t$. Welcome to calculus and differential equations. These won't be discussed here, but the concepts are actually not that hard (but the solutions can be). In practice, many of these equations have been solved for you, or a computer program does the work (see the Computer Modeling section later in this Appendix). Find some clever YouTuber and watch an introduction to calculus and differential equations. The video will open some doors of thinking that you might find useful.

Now let's look at some of the number crunching that goes on behind engineering and design programs so you can actually understand how values are being calculated.

Units

Engineering calculations in the United States often use the inch-pound-second (ips) system. This requires that the length be given in inches, the force in pounds, and the time in seconds. There is also the foot-pound-second (fps) system that uses the length in feet. These unit systems fall under the category of the *United States Customary System* (USCS or USC), or sometimes just called *United States customary units*. Using the terms *fps* and *ips*, or at least USCS, is a little more precise than just calling them *English units*. Note British Imperial units differ in their volume measurements, so be careful with the term *Imperial units*.

The metric system is also commonly used in engineering calculations in the U.S. and is standard outside the U.S. Almost all engineering textbooks use metric units. Specifically, this is referred to as the *Systeme International*, or by the less French description, *International System of Units*. The acronym is *SI* nevertheless.

Unit systems use base units and derived units. The base units are the foundational units of measurement, and other units derive from them. For example, the SI system has base measurements of distance, mass, and time. These base units are meters, kilograms, and seconds, respectively. The SI system's derived units are everything else, including units that you might be familiar with such as *newtons* (N) and *pascals* (Pa).

The U.S. system (USCS) has the same base measurements as the SI system, but there is some weirdness with the USCS ("English") mass unit. Sometimes to clarify that the pound unit (lb) is specifically for force and not mass, it will be called *pounds-force* (lbf). Mass can have various units such as *pounds-mass* (lbm) and *slugs*. One unique twist to this problem with the meaning of the unit *pound* is that the *gravitational ips* system uses the pound unit as force rather than mass, as is the case with the *absolute ips* system.

Within the metric system, there is a less common system, called the *cgs* system. The cgs system uses length in centimeters, mass in grams, and time in seconds. Force is derived from these and has the intriguing unit name *dynes*. This leads to the often asked question, "How many dynes in a pound?" The answer is 0.0000022481, which is pretty much the last thing in the world you should ever memorize. Now I am starting to wonder, how to convert dynes to slugs?

The electronics field uses a unit system derived from the SI system, but there are some different base units. These are called the *electromagnetic unit* (EMU) and *electrostatic unit* (ESU). Typically, we are only concerned with things like volts, amps, ohms, and watts. However, the EMU and ESU systems accommodate the full range of variables involved with electromagnetics.

The HVAC profession uses a mixed approach to units; we use metric units when talking about pulling a vacuum in *microns* and low-pressure measurements in the odd, English-derived unit of *inches of water column*. We often use acronyms as part of our jargon, such as *psi* (pounds per square inch), *cfm* (cubic feet per minute), *gpm* (gallons per minute), *BTU* (British thermal units), and *BTUH* (British thermal units per hour). But we have some units that those outside HVAC would not understand, such as *tons* (1 ton of cooling equals 12,000 BTUH), *inches* or *inches of water column* (IWC or sometimes just WC), *SCFM* (standard cubic feet per minute), and *microns* (1 millionth of a meter, but usually only used in evacuating refrigeration systems). Efficiency measurements also use a wide variety of industry-specific measurements and acronyms such as SEER, EER, HSPF, AFUE, COP, etc.

Thermodynamics

Ideal Gas Law

Let's start this dive into thermodynamics by looking at how pressure, volume, and temperature are related in gases. The ideal gas law puts this together, which looks like:

$$PV = nRT \tag{A.1}$$

Where:
P = pressure
V = volume
n = number of molecules
R = universal gas constant
T = temperature

This equation shows the relationship between pressure (P), volume (V), number of molecules (n), the universal gas constant (R), and temperature (T). The universal gas constant is just a conversion factor and does not change.

This relationship tells you many things. For example, let's consider constant volume systems, where V doesn't change. These are common in HVAC piping and ducting systems. In this case, the pressure and temperature change in the same direction. That is, if you increase the temperature, you increase the pressure. For example, when you drive your car, the tire pressure increases because the frictional heating of the tire causes the air temperature inside the tire to increase. Because the volume is constant, the only response of the closed, constant volume tire system is for the pressure to increase. When you measure your car's tire pressure after it has been driven, the pressure will always be higher, unless you have a leak and reduce the n value. Then you have other problems. The pressure in your tires goes up when you add air because the number of molecules (n) increases.

Of note, the ideal gas law applies only to low pressures such that (1) there are no forces between molecules and (2) molecules occupy negligible volume. The math gets altered a bit at high Mach speeds, although the relationship between the variables is the same.

First Law of Thermodynamics

Energy is conserved.

The change in heat is equal to the change in internal energy (temperature) plus the change in kinetic energy (velocity) plus the change in potential energy (usually pressure) plus the change in work (pumping energy or turbine extraction of energy).

This can be shown as follows:

$$\Delta Q = \Delta U + \Delta KE + \Delta PE + \Delta W \quad \text{(A.2)}$$

Where:
ΔQ = change in heat energy
ΔU = change in internal energy
ΔKE = change in kinetic energy
ΔPE = change in potential energy
ΔW = change in work energy

Energy is often manipulated between pressure and velocity, such as static and elevation pressure and the conversion of velocity to "velocity pressure" that is used in duct design. Friction takes away energy, and the odd thermodynamic term *work* either adds or subtracts energy. The Bernoulli equation, given below in Equation (A.3), provides the relationship between pressure and velocity when ignoring all the other stuff.

The Bernoulli equation calculates total pressure as follows:

$$P_{\text{total}} = P_1 + \rho V^2/2g_c \quad \text{(A.3)}$$

Where:
P_{total} = total pressure
P_1 = static pressure
ρ = fluid density
V = fluid velocity
g_c = gravitational constant

Second Law of Thermodynamics

The state of randomness (entropy) always tends to increase.

You can't throw individual parts of a clock at a brick wall and have them land as an assembled clock. But you can do the opposite—throw an assembled clock at a brick wall and watch what happens! Other definitions for the second law of thermodynamics you might hear include:

> *It is impossible to construct a device that operates in a cycle and produces no effect other than the transfer of heat from a cooler body to a hotter body.* This means that you can't construct a refrigerator that operates without an input of work.

> *A heated object will cool, but it will not absorb heat by conduction with air.* That is, some processes can only move in one direction—hot to cold.

Third Law of Thermodynamics

Entropy of a perfect crystal at absolute zero is zero.

This law provides a basis for comparing temperature and energy. It means that when a material is at absolute zero degrees (–460 °F), the atoms are perfectly frozen and aligned; they can't move or get disorganized. The need for this reference can be observed, for example, when calculating radiation heat transfer, which is presented later.

Heat Transfer

Note that all these heat transfer equations are for steady-state conditions as well as unchanging material properties, such as thermal conductivity. They also assume there is no internal heat generation. Transient heat transfer will be touched upon at the end of this Appendix when we look at computer programs that do thermal analysis.

As mentioned in the introduction of this Appendix, I'll use lowercase q for rate of heat transfer, as in BTU per hour. I'll use uppercase Q for total heat energy, as in BTU.

Conduction

Conductive heat transfer is material-to-material thermal energy movement. The bouncy molecules are sharing their thermal vibration with anything they are touching. For heat transfer applications, you want the heat transfer to be as high as possible. Therefore, you want a lot of surface area and high thermal conductivity, as can be achieved with copper and aluminum.

In building design, you generally want low thermal conductivity. Thermal load analysis programs allow you to put in construction details that calculate the total R value. While insulation provides most of the R value (resistance to conductivity), the programs will include the insulating ability of wood, siding, and other common building materials that are layered around the insulation.

The R value is the reciprocal of the U value that is commonly referred to in these calculations. That is,

$$U = 1/R \tag{A.4}$$

Where:
U = conduction resistance. This is the inverse of the R value.

The heat transfer due to conduction is calculated as:

$$q_{\text{conduction}} = UA\Delta T \tag{A.5}$$

Where:
$q_{\text{conduction}}$ = conductive heat transfer
A = area
ΔT = temperature change

Thermal contact resistance is another consideration in conductive heat transfer. Thermal contact resistance is a measure of the inefficiency of transferring heat between two surfaces due to the way the faces contact each other. For example, if a surface is rough, there are very few contact points and therefore high thermal contact resistance. Adding thermal grease and pressure reduces this resistance. Conversely, lightly touching surfaces with air gaps will have high thermal contact resistance.

Convection

Convection is caused by fluid movement. It is generally related to the fluid velocity squared. Other details, such as boundary layer thickness, go into this calculation also; however, this all gets buried in the mysterious term *film coefficient*.

Natural convection is produced by warmer air rising and dragging the surrounding cooler air with it. It's funny how the surface forces/stickiness of materials is so important in understanding their behavior. Unlike natural convection, the term *forced convection* means you are using a fan or pump to move fluid and increase convective heat transfer.

The heat transfer due to convection is calculated as:

$$q_{\text{convection}} = hA(T_s - T_{\text{inf}}) \tag{A.6}$$

Where:

$q_{\text{convection}}$ = convective heat transfer
h = film coefficient (h is a function V^2)
T_s = temperature at the surface
T_{inf} = temperature at infinite distance from the surface (ambient temperature)

Radiation

Radiation is produced when surface temperatures are above absolute zero (–460 °F). Surfaces radiate to each other, with the net radiation shown below in Equation (A.7).

Thermal load analysis programs will allow you to enter the type of roofing and windows, as well as many other details that calculate radiation. For commercial buildings, the absorption of solar energy is a big thermal load.

A couple of special notes on radiation: The radiation heat transfer is proportional to the solid angle it makes with the other surface. *Mean radiant temperature*

(MRT) is a term often used, and it is the summation of angles multiplied by temperatures. For example, if you have radiant floor heating, the wall and the ceilings are also part of the mean radiant temperature.

Below is the equation for heat transfer due to radiation for a surface radiating to an ambient, uniform, surrounding surface.

$$q_{\text{radiation}} = E A S (T_{s1}^{4} - T_{s2}^{4}) \tag{A.7}$$

Where:

$q_{\text{radiation}}$ = radiant heat transfer
E = emissivity
A = surface area
S = Stefan-Boltzmann constant
T_{s1} = temperature at surface 1
T_{s2} = temperature at surface 2

Surface-to-surface radiation has a more complex equation, which I won't include here, but it includes the variable called *view factor*, commonly called *F*. It is a dimensionless number that takes into account the relative orientation, distance, and shape of radiating and absorbing surfaces.

Internal Loads

This is usually the biggest thermal load category in commercial applications. Thermal load analysis programs allow you to enter internal loads produced by people and equipment. People and things like plants create both *sensible* and *latent* heat. Sensible heat is what you feel (sense, get it?), and latent heat is humidity. The breakdown of sensible and latent heat (temperature and humidity) is important for specifying air handler cooling coils, among other things.

When people exhale, they immediately add latent heat. However, below is an interesting calculation that considers the time delay from when warm people radiate to cool walls and heat them up and the walls reradiate the heat back into the room. The heat transfer produced by people is a function of the number of people, how much heat they produce, and a delay factor as shown in Equation (A.8).

$$q_{\text{people}} = N(\text{SHG})(\text{CLF}) \tag{A.8}$$

Where:

q_{people} = heat transfer from people
N = number of people
SHG = sensible heat gain per person
CLF = cooling load factor. Typically, it starts at 0.6 to 0.75 for the first hour. CLF accounts for radiation-to-reradiation delays.

Evaporation

Energy is removed from a fluid when the high-energy molecules move from a liquid to a gas phase. This removal of the highest-energy molecules lowers the temperature of the remaining fluid. There are various formulas for calculating how quickly water evaporates, but the energy loss due to evaporation is:

$$q_{\text{evaporation}} = m\,\Delta H_v \tag{A.9}$$

Where:

$q_{\text{evaporation}}$ = heat transfer from evaporation
m = mass
ΔH_v = heat of vaporization. This is the energy required to change a liquid to a gas phase.

Infiltration

Infiltration is the uncontrolled entrance of air into a conditioned space. *Controlled ventilation* is when you pull in outside air to reduce carbon dioxide levels in the supply air. For neutrally pressurized buildings, the infiltration is equal to exfiltration. *Exfiltration* is the uncontrolled exiting of air from a conditioned space. The air pressurization is controlled by the air handler, which controls the flow of the regulated outside and exhaust air. Infiltration sneaks around the air handler's control.

Thermal load analysis programs let you enter the type of infiltration issues, such as barriers and windows, that allow for infiltration calculation. The calculation is based on air changes per hour (ACH) and the condition of the air that is infiltrating.

A common calculation for heat exchange in a room due to infiltration is:

$$q_{\text{infiltration}} = V \Delta T (\text{ACH})\, k \tag{A.10}$$

Where:

$q_{\text{infiltration}}$ = heat transfer from infiltration
V = room volume
ΔT = temperature change
ACH = air changes per hour
k = conversion factor (In English units it would be 0.018 BTU/°F ft^3.)

Thermal Storage

Thermal storage is used to store energy for when it's needed, whether it's for heating or cooling. Water has a great ability to store heat with a small temperature difference because its specific heat is high. The total energy stored is typically designated with a capital Q rather than the lower case q that indicates the rate of heat transfer. Total energy is measured in BTU while rate of heat transfer is measured in BTU/hr. The calculation for the total energy stored is simply the temperature difference multiplied by the thermal mass:

$$Q_{\text{stored}} = \rho\, V c\, \Delta T$$

Where:

ρ = density
V = volume
c = specific heat (water = 1)
ΔT = temperature change

or

$$Q_{\text{stored}} = m\, c\, \Delta T \tag{A.12}$$

Where:

m = mass

Pipe Friction and Head Loss

The Hazen-Williams equation is a good general formula for friction loss in pipes and can be used in conjunction with equivalent length calculations for fittings. It's empirically derived, meaning it is based on reality and not just theory. Also, these equations include English units, so it's more than just the relationships shown in previous equations—you can easily calculate pressure drop values with these equations.

The Hazen-Williams equation is commonly used because it is easier than the Darcy-Weisbach approach, which is better over a wider range of flows. The Darcy-Weisbach equation is valid for both laminar and turbulent flows and works with the Moody diagram or Colebrook equation. The Moody diagram is interesting if you spend time with it. It shows the changes in friction as a function of laminar versus turbulent flow as well as the effect, and lack of effect, of pipe roughness. It's something you may want to check out sometime—there's a lot of information given in this intimidating family of curves.

The Hazen-Williams formula is valid for turbulent flow at "reasonable" velocities. The nonintegral exponents are very humbling; they sort of show who wins the practice versus theory battle. When I actually size pipes and pressure drops, I use both equations and check against an old-fashioned table to make sure they are all in the same range.

Note that both the Hazen-Williams and Darcy-Weisbach equations only work when the pipe is completely full. If you are dealing with open channel flow, which is when a pipe or channel is not completely full of water, such as with drains, then you need another approach. The Manning equation is a similar empirically derived equation that can be used for calculating velocity in open channel flow.

Hazen-Williams Equation for Pressure Loss in Pipe

The fluid pressure drop per 100 feet of pipe is calculated as:

$$\Delta P_{f100} = 452\,(Q^{1.85}/C^{1.85}d^{4.86}) \tag{A.13}$$

Where:
ΔP_{f100} = pressure drop per 100 feet of pipe in psi (lb/in^2)
Q = flow in gallons per minute (gpm) (Q is used here, which is sort of confusing because Q is normally a reserved variable name for heat energy. Sorry, I don't make this stuff up.)
C = friction constant (The C value varies with surface roughness. It is a unitless value. Some approximate C values are copper pipe equals 135 and steel pipe equals 100.)
d = pipe diameter in inches (in)

Head Loss

The pressure change due to lifting a fluid is calculated as:

$$\Delta P_g = \rho \Delta h/144 \tag{A.14}$$

Where:
ΔP_g = pressure drop due to weight of fluid in psi (lb/in^2)
ρ = density in lb/ft^3
Δh = vertical height in feet (ft)

Fans

Moving air is important, and a lot of fans are packaged into air handlers, air-cooled chillers, etc. However, the following ideas may be helpful, especially when considering connecting air handlers to plenums.

Fan Laws

You may have heard of these fan laws. You might find them helpful, so here they are:

1. Airflow is proportional to fan speed.
2. Static pressure varies directly with the square of the airflow.
3. Power varies directly with the cube of the airflow.

And that is all I have to say about that.

Maybe more useful is to note what happens at the other end of the fan output, that is, how do people react to air movement. Air movement under 50 feet per minute (fpm) is unnoticed, 50–250 fpm is noticed (but pleasant), and over 300 fpm feels drafty.

Fan Outlet

Fan connections are important for efficiency. For $V_o > 2{,}500$ fpm, the calculation for the effective duct length for the discharge of a fan into a plenum so that it provides fully developed flow is:

$$L_e = V_o (A_o^{0.5})/10{,}600 \qquad \text{(A.15)}$$

Where:
L_e = effective duct length in feet (ft)
V_o = duct velocity in feet per minute (fpm)
A_o = duct area (in^2)

For $V_o < 2{,}500$ fpm, the calculation is:

$$L_e = (A_o^{0.5})/4.3 \qquad \text{(A.16)}$$

Fan Inlet

A general guide for the air inlet is to have a straight transition equal to 6 times the diameter (or the equivalent diameter for a rectangular duct) of the return air duct.

Atomization in Nozzles

Nozzles are used to atomize fuel for combustion and water in cooling towers. When a fast stream of liquid travels out of an orifice, a field of droplets and induced air develops outside the fluid core in the center. Droplets form as a result of a combination of turbulence, aerodynamic interactions, and surface tension. The droplet size and the time required for them to form are characterized by the *Weber number*. This is a dimensionless parameter like the Reynolds number and Mach number we had considered earlier when looking at fluid flow. The Weber number is the ratio of inertia forces and the surface tension. If you want a liquid to atomize, it is easiest if it flows fast and has a weak surface tension. Water takes a lot of velocity to atomize because it has strong surface tension!

The calculation for the Weber number is:

$$We = \rho^2 V^2 d/\gamma \tag{A.17}$$

Where:
We = Weber number
ρ = gas phase density
V = relative velocity of the droplet to the gas velocity
d = droplet diameter
γ = surface tension

Nozzles can have a tough life, especially with certain fuels. Nozzle orifices are degraded by abrasive wear and surface fatigue. Abrasive wear is caused by the plowing of two materials in which a material is then deformed. In surface fatigue, the material is reduced in strength due to cyclic (on and off) stresses that initiate and propagate cracks. These cracks grow large enough to cause flaking of the surface material. The wearing on the orifice can also be caused by *tribochemical wear*. Tribochemical wear is a combination of mechanical and thermal processes at the contact interface, which changes the surfaces' material properties.

If there are solids in your fuels, the solid materials traveling through the atomizer will produce a lot of friction. I have dealt with this problem with some alternative fuels, but it is not a common concern in most HVAC systems. However, if solid particles are in the fuel, the friction produced between particles occurs where the peaks and valleys of the particle surfaces interface. This contact area is much smaller than would be provided by an apparent contact area of a perfectly smooth surface. Therefore, since the area is so small (or highly *localized* in engineer-speak), the loads are exaggerated. You can get *spalling* caused by local stress, called *tri-axial stress*, that overwhelms the material. With further load however, the surface roughness gets compacted and the contact area increases. The rough surfaces deform because of the large contact area.

Thermal Expansion

Material that is heated or cooled will either expand or contract. If the objects are rigidly anchored, the expansion effect is turned into internal stresses. The equation for thermal expansion is simply:

$$L_2 = L_1 \propto \Delta T \tag{A.18}$$

Where:
L = length
$\propto$ = coefficient of thermal expansion
ΔT = temperature change

Table A.1 shows the coefficients of linear thermal expansion for common materials.

TABLE A.1 Coefficients of Thermal Expansion for Common Pipe Materials

Material	in/in/°F*
Carbon steel	0.0000065
Copper	0.0000093
Stainless steel	0.0000095
Aluminum	0.000013
PVC	0.000028
CPVC	0.000044
HDPE	0.000067
PE	0.000083

*The exact value depends on the specific alloy or plastic composition.

If internal stresses develop, they are connected with the equation that relates stretchiness (strain) to stress. This equation is:

$$\sigma = E\,\varepsilon \tag{A.19}$$

Where:
σ = stress
E = Young's modulus (or *modulus of elasticity*, which is a broader term)
ε = strain

If you combine these two equations, you get this thermal stress equation:

$$\sigma = E \propto \Delta T \tag{A.20}$$

Where:
$\propto$ = coefficient of thermal expansion
ΔT = temperature change

Material Strength

The load that material can handle is a function primarily of the cross-sectional shape and the force applied. If force is applied over a long lever arm, this increases the twisting action, which amplifies the stress. The twisting action that you are familiar with is technically called the *moment* or *torque*. The calculation for the moment of a simple cantilever beam is simply force times distance. You're familiar with breaker bars and the wonderful things they can do, namely increasing the moment. You also know what a sledgehammer blow on a well-made wrench can do by amplifying the force. You increase the moment by using a long lever (or moment arm) and a big force. Moment is commonly measured in the descriptive units of inch-pounds (in-lbs) or foot-pounds (ft-lbs).

The moment, or torque, that a car engine produces is made by using a large cylinder that can handle a lot of fuel and create a big explosive force along with a crankshaft with a big offset. The large cylinder is the force, and the crankshaft offset is the lever. Multiplying the force of the cylinder explosion (in pounds) by the offset distance of this cylinder (in inches) gives you torque (in inch-pounds). Horsepower is simply a measure of how fast torque is applied. To maximize horsepower, you want high crankshaft revolutions with big or multiple cylinders acting on a big offset crankshaft. You can get big horsepower, but lower torque, with smaller cylinders run at high revolutions. The converse is true; torquey engines might not be able to run very fast, but that's what transmissions are for.

Back to the topic, the equation for moment for a simple cantilever beam that is anchored at one end and has a force at the other is:

$$M = Fd \tag{A.21}$$

Where:
M = moment
F = force applied
d = distance between force and anchor

If the load is supported at both ends, this reduces radically to one-fourth the value of the cantilever beam, or:

$$M = Fd/4 \tag{A.22}$$

The stiffness and strength of the cross section is defined by something called the *moment of inertia*. The letter I is reserved for this term and is measured in an odd unit of inches to the fourth power, in^4.

For a rectangular cross section, the equation is:

$$I = bh^3/12 \tag{A.23}$$

Where:
I = moment of inertia
b = thickness of rectangular cross section
h = height of rectangular cross section

For a round cross section, the calculation for the moment of inertia is:

$$I = \pi r^4/4 \tag{A.24}$$

Where:
r = radius of a round cross section

Note that the height of a material increases the moment of inertia as a cubic function. If you double the height of a beam, it becomes eight times stronger. If you double the width of the same beam, it only becomes twice as strong. You know this sort of thing from practice, but it's interesting to see it in a mathematical presentation. Knowing where to put material is important in trying to make things cheaper and lighter. Overdesigning things is easy.

Finally, you can calculate the stress on a material and compare that value with what the material can actually handle. The stress value is calculated in pounds per square inch (lbs/in^2 or psi). For simple tensile loading, like a pipe hanger, the equation is stress equals force divided by area, or:

$$\sigma_t = F/A \tag{A.25}$$

Where:
σ_t = tensile stress
F = force applied
A = cross-sectional area

Flexural loading is the most common loading in most mechanical systems outside of hangers. The stress calculation for this is uglier, but it has really interesting details:

$$\sigma_f = Mc/I \tag{A.26}$$

Where:
σ_f = flexural stress
M = moment
c = distance from the neutral axes (usually the center) to the outermost edge
I = moment of inertia

For a rectangular cross section, the maximum stress is when c is equal to half the height. Notice, if c equals zero, which it does in the middle of a beam, then the stress calculates to be equal to zero. That is why you need to drill holes in the middle of beams! This equation also tells you that a big moment of inertia value gives you low stress, while a big moment (not moment of inertia) produces a high stress. Take a moment and think about that.

If the stress is below the yield stress of a material, it won't usually break if it leads a peaceful life. However, fatigue, impact, corrosion, and other things can make something break outside this mathematical calculation. The worst deviation from this math is buckling, which can occur well below the yield stress on long slender objects. In practice, you will probably find fatigue failure and corrosion to be the most commonly occurring unanticipated way by which things break.

Figure A.1 illustrates how shape affects mechanical behavior. This diagram shows the behavior of I-beams versus pipes in terms of their bending and torsional behaviors. Our bones are like tubes because their response to load is predictable and equal in all planes (plus the inside can be used to store and run important stuff like bone marrow, nerves, and blood vessels). You use I-beams only when you have predictable loads (and gravity!).

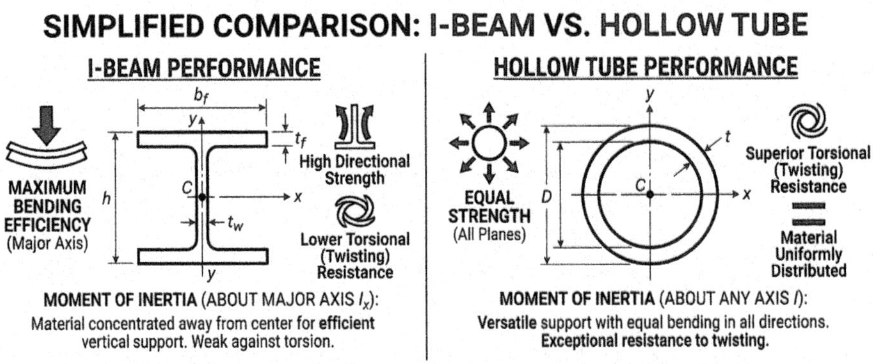

FIGURE A.1 Comparison of shape and strength characteristics. Note that the *h* value of the I-beam and the *D* value of the tube are the most important contributors to strength and rigidity.

This discussion of material strength is not meant to be a review of structural engineering. Not everything is covered here. However, you know how multidisciplinary HVAC is. You also work with heavy, vibrating machinery that can cause structural problems. This introduction provides some insights into the math connected with material strength and gives you something to lean on.

Fracture Face

The area where something breaks is called the *fracture face*. The fracture face can be an indicator of how something breaks, because there are distinctive differences in how something breaks when it is under impact, torsion, flexural, tensile, compressive, or shear loading. Fatigue failure has a distinctive fracture face with many unique features such as beachmarks, as shown in Figure A.2.

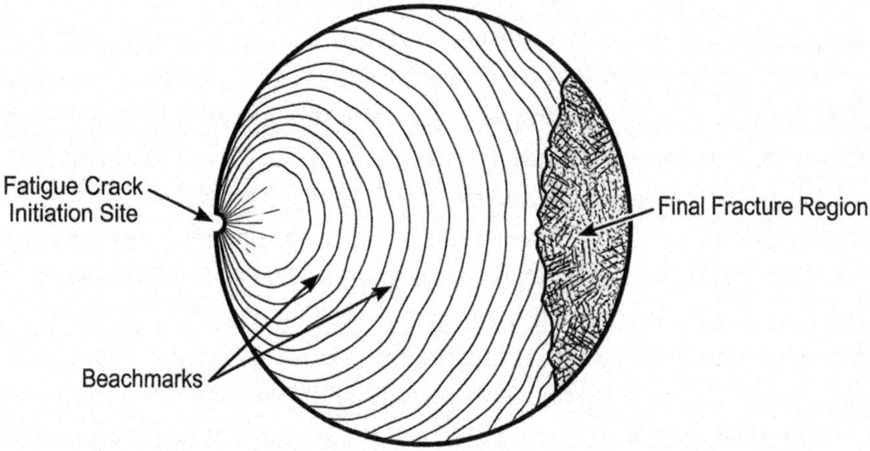

FIGURE A.2 Fracture face characteristics of a shaft failing due to fatigue.

Brittle and ductile materials also present different fracture faces. Figure A.3 shows an example of the fracture face from a torsional overloading.

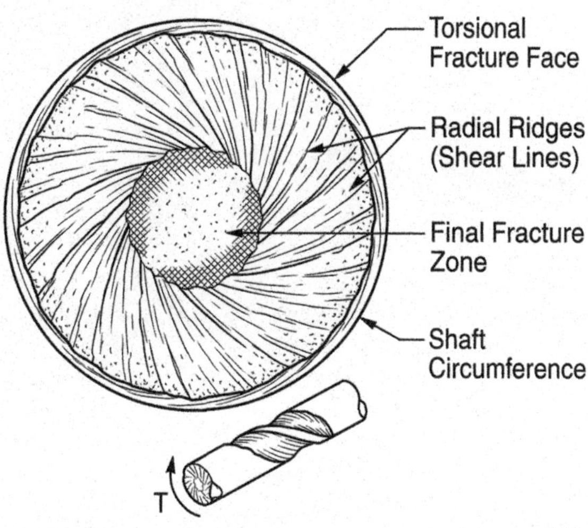

FIGURE A.3 Fracture face characteristics of a shaft failing due to torsional overload.

FORENSICS: WHEN THE FRACTURE TELLS A STORY

I was asked to inspect a new but inoperative chiller in connection with a warranty claim. I approached the freshly painted, robust-looking chiller and was amazed at what I found. When I removed the shiny new access panels (held in position by a shiny new frame), I saw that the guts of this chiller were damaged. This chiller was new and had just been delivered; however, broken pieces, scratched paint, and other evidence indicated the torrid story of this poor chiller. One of the refrigerant lines was broken, solenoid valve coils were disconnected, and there were many scratches and other damage that were inconsistent with the otherwise excellent condition of the chiller. Inspecting the fracture face of the broken copper refrigerant line (Figure A.4) showed burrs protruding both inwardly and outwardly along the entire circumference of the refrigerant line on both the sides of the fracture. The appearance of the copper would suggest impacting.

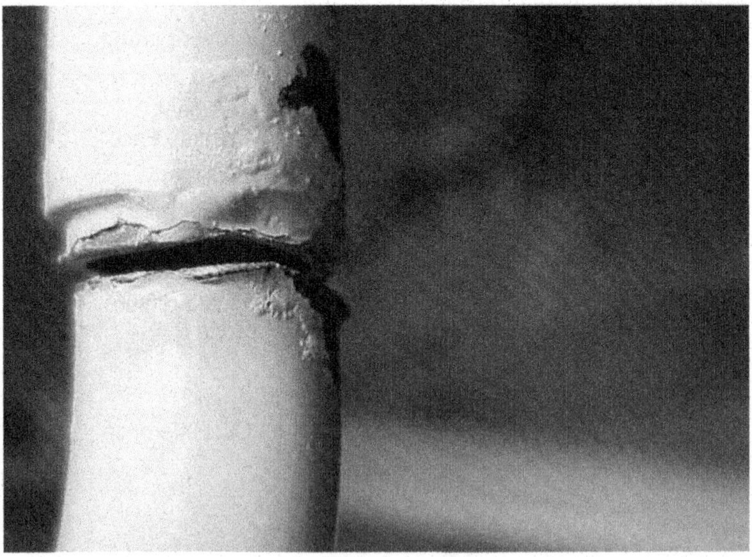

FIGURE A.4 Fracture face of refrigerant tubing that had burrs on the outside and inside.

I was confused about how the internals of the chiller could have been damaged; the refrigerant line was broken, and yet the chiller looked brand new. When I mulled over all the evidence, the only conclusion I could come to was that the chiller had rolled over, thereby suffering damage, including a broken refrigerant line. After the rollover, some features were repainted, and a new framework and panels were attached to the broken chiller. Then it was shipped. The fracture faces of the broken copper refrigerant line, I believe, impacted each other during shipping to produce the unusual burring. That was the conclusion I provided the insurance company.

Computer Modeling

Computer programs can accurately determine the thermal loads for houses and buildings as well as the size of equipment, pipes, and ducts. However, I want to introduce some computer modeling ideas that might be interesting and tie in how the equations we considered are used in numerical modeling.

I've seen programs go from cluttered, difficult interfaces to having the user simply draw the plan view of a house or building in an easy-to-use interface and then clicking options. The thermal load calculation programs have accurate climatic data, insulation values of common construction practices, great duct loss analysis, and much more.

Not only can programs do thermal load analysis, but they can also size equipment, pipes, and ducts, as well as generate bills of materials. And the list of capabilities could go on. These programs are amazing time-savers. On top of analysis software are building information modeling (BIM) systems that coordinate all the building information into one holistic package.

For numerical modeling, let's consider two powerful programs: finite element analysis (FEA) and computational fluid dynamics (CFD). These programs use many of the equations presented previously. They take a physical thing and turn it into a mesh of nodes and elements. The nodes mark three-dimensional space, and the elements mark the relationship between the nodes. The elements store the relevant math equations. The nodes are changed by how the equations behave. The nodes can be moved, such as in the case of a force being applied to them; or they can heat up, such as when a temperature is applied to them. Of course, the nodes would also move because of thermal expansion—it's all intertwined.

The elements connecting and driving the nodes are equations. For example, FEA will calculate the temperature and rate of change of temperature in an object by breaking it up into little pieces and using the thermal conductivity between the nodes based on the material selected by the engineer. In this way it maps out the temperature characteristics of the object. It can also consider radiant heat transfer and convection if the convection coefficient is clearly obtained.

CFD analyzes fluid flow. Like FEA, CFD involves the reduction of complex shapes into simple elements for which solutions can be obtained. An object under investigation is meshed or divided into numerous infinitesimally small nodes. The nodes are then connected by elements that contain material information that dictates how the adjoining nodes will behave. CFD looks specifically at fluid flow and relies on some form of Navier-Stokes equations. These are partial differential equations that describe the motion of fluids. Differential equations are used in all these numerical methods because they can handle transient, time-based heat transfer. They can also model changes in variables, such as temperature differences and thermal conductivity. CFD does a better job of handling convection than FEA, unless the convection coefficient is known and is unchanging.

Be careful when looking at the results from these programs. The vibrant colors and apparent complexity of FEA and CFD infer accuracy even when the results are wildly incorrect. Inaccuracies can arise from many things, including incorrect boundary conditions; incorrect application of external factors; inadequate mesh density; and, of course, incorrect rendering of the parametric model. Garbage in, garbage out applies to these computing beasts also.

Electric Power

Electric power is equal to voltage multiplied by current. This calculation is:

$$P = EI \tag{A.27}$$

Where:
P = power (watts)
E = voltage (electromotive force, volts)
I = current (amps)

Power loss in wiring is a function of the resistance of the wire and the current squared. This derives from Ohm's law, which states voltage is equal to current multiplied by resistance ($E = IR$). Algebraic mixing and manipulation of this with $P = EI$ gives:

$$P = I^2R \tag{A.28}$$

Where:
P = power (watts)
I = current (amps)
R = resistance (ohms)

If you operate a system at lower voltage, it requires more current because of the whole $P = EI$ formula thing, which would look a lot better as $P = IE$ (I'll offer as a personal note, I prefer apple). Because you have to use more current to get the same power, the resistance loss goes way up because the power loss is mathematically related to the current squared.

High voltage means high electrical efficiency. You can use smaller wires for the same voltage drop. In addition, flowing current produces electromagnetic waves, like a radio transmitter, and produces a distinctive resistance called *reactance*. Reactance is a measure of how wires, like in a motor coil, can act like a capacitor and inductor and resist current as a function of frequency.

Capacitors have low reactance to incoming high frequencies and high reactance to incoming low frequencies. Inductors behave in the opposite way. Inductors like to let low frequencies go through but resist high frequencies. Therefore, electrical wiring and loads are often based on the odd unit of VA, which stands for *volt-amp*, as opposed to using the term *watts*. This VA value is called *apparent power*. Apparent power is only a serious issue for big equipment, where you are dealing with thousands of watts, and so most commonly you see units of kVA and kW.

The reason for the difference between apparent power (measured in kVA) and real working power (measured in kW) is that in alternating current systems involving lots of wire, like an electric motor, there is a large reactance. Reactance inhibits alternating current and produces a phase difference between the voltage and current, called the *power factor*. The power factor is the ratio of power that actually does work as compared with the apparent power. The power factor ranges from −1 to +1. The power factor will be 1 if the power is not at all capacitive or inductive—that is, there is no reactance.

The formula for apparent power (S) is:

$$S = P/PF \tag{A.29}$$

Where:
S = apparent power (kVA)
P = power (kW)
PF = power factor

Therefore, a motor's power, in kW, dictates how big a pump or fan it can drive, but kVA determines the wire sizing and heat produced. A low power factor means electric power is being used inefficiently. This is a very important issue in commercial HVAC because consumption of electricity is a huge operating expense.

For completeness, there is a measure that combines reactance and good, old-fashioned electrical resistance. It's called *impedance* and is important for matching electrical devices. I talked about this in Chapter 13, "Electronics: Building Automation, Communication, and Motors." You could argue that we should always talk about impedance rather than just reactance, but I just go with whatever the big motor people prefer.

Ethics

I worked on a concrete crew in high school. When I first went to put a keyway into the top of a just-poured footer using a 2×2 square keyway shape, my boss stopped me and said to just drag the edge of a 2×4 lumber along the footer. I did what he said. I saw my sad little V-shaped keyway and remembered small stones falling into it, making it function even worse.

When I was an undergraduate work-study student, we had to have a form signed every semester. Because we worked all over the country, there was no way the form could get routed from our home address to our work address in time. Our college dean told us that the university did not have handwriting experts checking our signatures. It was one of my first experiences with situations where the sacrosanct notion of signing a document was ridiculed and signature forgery was suggested.

A young person's standards get challenged by someone in authority, and the young person's standards shift in response.

You've heard people joking about *de minimis* refrigerant gas release. You've seen heat tracing on 24/7/365. You've seen potable water leaking endlessly. You have probably seen a lot of bad job installations. When I bought my current house, the attic air handler's condensate line ran to the stack vent. To add to the indignity, the PVC stack was opened up with a bashed-in hole from a hammer strike. I could go on, but I had better stop. However, you have probably seen what a reciprocating saw does to joists and headers as people try to run pipes and ducts through things.

These are some of the ethical challenges you deal with day to day. You have probably dealt with much bigger ones too, I know I have, and have lost much sleep because of them. It's a messy world.

The legal system is generally good at not criminalizing honest mistakes. They happen. Even large groups of experts make mistakes. Many years ago, I was on an American Gas Association technical committee, and we all missed the potential problem with thin-wall CSST tubing and lightning strikes. At that time, we were more worried about accidental nail penetrations and the potential problems with outdoor use because of pitting corrosion and the effect of muriatic acid washes. No one thought about what would happen if a large current flow was induced into the tubing.

Slowly our standards are shifted by the culture. The lengthy, indecipherable legal language we are compelled to agree with also seems to weaken our ability to understand all the things that we lay our formal signature upon. To add to the degradation, you can just click a mouse button to assert we comply with a bunch of indecipherable words on a screen.

Morals come from a greater source than a list of ethics produced by people sipping coffee in a conference room. However, ethics can often have some legal foundations that are often, but not always, captured by regulations. These regulations can be annoying, but they are usually rooted in something that ensures public safety. You have probably earned your EPA 608 certification among other credentials, which ensures the public of a certain level of competence it should expect from you.

Externally imposed ethical doctrines can be referred to as *professional ethics* to clarify them from the generic term *ethics*. Professionals can be considered liable if they are negligent in their work. It is helpful to recognize your legal standing as a professional who "professes" expertise in something. This summary of the standard to which you might be held comes from the 1954 State of California's case law in *Gagne v. Bertran* (43 Cal.2d 481). Under "The Cause of Action for Negligence," it states:

> The services of experts are sought because of their special skill. They have a duty to exercise the ordinary skill and competence of members of their profession, and a failure to discharge that duty will subject them to liability for negligence. Those who hire such persons are not justified in expecting infallibility, but can expect only reasonable care and competence. They purchase service, not insurance.*

There are some vague sources to lean on for ethical guidance. For example, the National Society of Professional Engineers (NSPE), in its Code of Ethics, states that engineers must "[h]old paramount the safety, health, and welfare of the public."† However, this isn't a statement you can hold in front of your boss or client and have the person automatically back down.

Ethical dilemmas have been the most personally challenging aspect of my career. I have had several ethical challenges that have given me much grief and tangible losses. My worst ethical problems have not arisen while employed at big corporations or small companies, but rather at middle-sized companies. These companies lie below the radar of contingency lawyers yet don't have the personal moral connection with a sole proprietor. However, I have had ethical challenges with all sizes of companies. In my personal experience, most companies have withdrawn from unethical actions when confronted.

* SCOCAL, *Gagne v. Bertran*, 43 Cal.2d 481, Stanford Law School, https://scocal.stanford.edu/opinion/gagne-v-bertran-26623.

†NSPE Code of Ethics, National Society of Professional Engineers (NSPE), www.nspe.org/resources/ethics/code-ethics.

Safety

Let's end this Appendix with a concise summary of safety concerns in HVAC. This is not a complete list; your list is probably a lot longer. However, I'm sure you have had plenty of training and enjoy good common sense, so this list is just a humble summary of critical, less obvious concerns:

- HVAC equipment is heavy; check load-bearing ability of structures and floors.
- Corrosion and fatigue are the leading causes of component failures.
- Things expand when heated and contract when cooled.
- Poor combustion results in undetectable carbon monoxide.
- Fuel can explode at certain fuel-to-air ratios.
- High surge pressure is produced by the closure of fast-acting valves.
- Water expands as it freezes, which can tear apart systems.
- Smoke travels through ducts.
- Refrigerants are heavier than air and can suffocate people.
- Steam and high voltage are predatory.

ACKNOWLEDGMENTS

I wish to thank all those that reviewed this book for ensuring its accuracy and appropriate presentation. I would like to especially thank faculty members Jason Killinger, Bradley Kishbaugh, Todd Woodling, and former students Mark Boone, Seth Carberry, Nick Pappas, Matt Robinson, Jesse Swartzlander, and Tyler Macdonough. Wrapped in those names is a huge amount of intelligent insights and practical experience that I was able to lean into. Thanks also to Judy Bass at Industrial Press for her enthusiastic encouragement, professional excellence, and careful treatment of this book.

INDEX

References to figures and tables are in italics.

ABOUT THE AUTHOR

Thomas Ask spent much of his career as a practicing engineer and now he teaches at Penn College. He earned his bachelor's degree in mechanical engineering from the University of Illinois, doctorate in industrial design from Middlesex University and is a licensed Professional Engineer. He has worked as a Research Engineer at the University of Illinois, Senior Engineer at Ingersoll Rand, Vice President of Engineering at Odin Systems International, and Principal of Ask and Associates. Tom has designed many products and systems and supervised testing with UL and FM. His designs include products and systems for combustion, natural gas piping, fire protection, and custom engineered systems. Tom has a "hands on" background and has earned his NATE Service Technician certification for air conditioning and heat pumps, EPA608 Universal Certification, and a NOCTI Plumbing instructor rating. Tom was a faculty advisor of ASHRAE and served on technical committees for AGA and SAE.